P9-CKU-259

574.92
N213

Understanding
MARINE
BIODIVERSITY

A Research Agenda for the Nation

WITHDRAW

Committee on Biological Diversity in Marine Systems

Ocean Studies Board
Commission on Geosciences, Environment, and Resources

Board on Biology
Commission on Life Sciences

National Research Council

LIBRARY ST. MARY'S COLLEGE

NATIONAL ACADEMY PRESS
Washington, D.C.   1995

NATIONAL ACADEMY PRESS • 2101 Constitution Ave., N.W. • Washington, DC 20418

NOTICE: The project that is the subject of this report was approved by the Governing Board of the National Research Council, whose members are drawn from the councils of the National Academy of Sciences, the National Academy of Engineering, and the Institute of Medicine. The members of the committee responsible for the report were chosen for their special competence and with regard for appropriate balance.

This report has been reviewed by a group other than the authors according to procedures approved by a Report Review Committee consisting of members of the National Academy of Sciences, the National Academy of Engineering, and the Institute of Medicine.

The work was sponsored by the U.S. Department of Energy under Grant No. DE-FG02-94ER61738, National Science Foundation's Grant No. OCE-9304519 with a contribution from the National Oceanic and Atmospheric Administration, Office of Naval Research's Grant No. N00014-94-1-0526, Department of the Interior, National Biological Survey's Grant No. 14-45-0009-94-1200.

Library of Congress Cataloging-in-Publication Data

Understanding marine biodiversity : a research agenda for the nation /
   Committee on Biological Diversity in Marine Systems.
       p.   cm.
   "The workshop, entitled Biological Diversity in Marine Systems,
was held May 24-26, 1994, at the Arnold & Mabel Beckman Center of
the National Academy of Sciences (NAS) in Irvine, California"—Pref.
   "Ocean Studies Board, Commission on Geosciences, Environment, and
Resources; Board on Biology, Commission of Life Sciences, National
Research Council."
   Includes bibliographical references (p.   ) and index.
   1. Marine biology—Research—Congresses. 2. Biological diversity—
Research—Congresses. I. National Research Council (U.S.).
Committee on Biological Diversity in Marine Systems. II. National
Research Council (U.S.). Ocean Studies Board. III. National
Research Council (U.S.). Board on Biology.
   QH91.A1U485   1995
   574.92′072—dc20   94-44420

International Standard Book Number 0-309-05225-4

Copyright 1995 by the National Academy of Sciences. All rights reserved.

COVER ART: "Marine Mardi Gras"© 1994 Barbara Wallace. While growing up on a cattle ranch in rural Trout Lake, Washington, Barbara Wallace developed her love of animals and nature, which has become the subject of her work. Recently she and her family relocated to a small farm in Trout Lake, where they are happily breeding horses and collecting animals of all kinds.

Printed in the United States of America

COMMITTEE ON BIOLOGICAL DIVERSITY IN MARINE SYSTEMS

CHERYL ANN BUTMAN, *Co-Chair*, Woods Hole Oceanographic Institution, Massachusetts
JAMES T. CARLTON, *Co-Chair*, Williams College - Mystic Seaport, Connecticut
GEORGE W. BOEHLERT, National Marine Fisheries Service, Monterey, California
SUSAN H. BRAWLEY, University of Maine, Orono
EDWARD F. DeLONG, University of California, Santa Barbara
J. FREDERICK GRASSLE, Rutgers University, New Brunswick, New Jersey
JEREMY B.C. JACKSON, Smithsonian Tropical Research Institute, Panama
SIMON A. LEVIN, Princeton University, New Jersey
ARTHUR R. M. NOWELL, University of Washington, Seattle
ROBERT T. PAINE, University of Washington, Seattle
STEPHEN R. PALUMBI, University of Hawaii, Honolulu
GEERAT J. VERMEIJ, University of California, Davis
LES WATLING, University of Maine, Orono

Staff

MORGAN GOPNIK, Ocean Studies Board, *Study Director as of 5/94*
DAVID WILMOT, Ocean Studies Board, *Study Director until 5/94*
MARY HOPE KATSOUROS, Ocean Studies Board, *Director*
ERIC FISCHER, Board on Biology, *Director*
LaVONCYÉ MALLORY, Ocean Studies Board, *Project Assistant*

OCEAN STUDIES BOARD

WILLIAM MERRELL, *Chair*, Texas A&M University, Galveston
ROBERT A. BERNER, Yale University, New Haven, Connecticut
DONALD F. BOESCH, University of Maryland, College Park
KENNETH BRINK, Woods Hole Oceanographic Institution, Massachusetts
GERALD CANN, Independent Consultant, Rockville, Maryland
ROBERT CANNON, Stanford University, California
BILIANA CICIN-SAIN, University of Delaware, Newark
WILLIAM CURRY, Woods Hole Oceanographic Institution, Massachusetts
RANA FINE, University of Miami, Florida
JOHN E. FLIPSE, Texas A&M University, Galveston
MICHAEL FREILICH, Oregon State University, Corvallis
GORDON GREVE, Amoco Production Company, Houston, Texas
ROBERT KNOX, Scripps Institution of Oceanography, La Jolla, California
ARTHUR R. M. NOWELL, University of Washington, Seattle
PETER RHINES, University of Washington, Seattle
FRANK RICHTER, University of Chicago, Illinois
BRIAN ROTHSCHILD, University of Maryland, Solomons
THOMAS ROYER, University of Alaska, Fairbanks
LYNDA SHAPIRO, University of Oregon, Charleston
SHARON SMITH, University of Miami, Florida
PAUL STOFFA, University of Texas, Austin

Staff

MARY HOPE KATSOUROS, *Director*
EDWARD R. URBAN, JR., *Staff Officer*
ROBIN PEUSER, *Research Associate*
MARY PECHACEK, *Administrative Associate*
LaVONCYÉ MALLORY, *Senior Secretary*
CURTIS TAYLOR, *Office Assistant*
ROBIN ALLEN, *Senior Project Assistant*

COMMISSION ON GEOSCIENCES, ENVIRONMENT, AND RESOURCES

M. GORDON WOLMAN, *Chair*, Johns Hopkins University, Baltimore, Maryland
PATRICK R. ATKINS, Aluminum Company of America, Pittsburgh, Pennsylvania
EDITH BROWN WEISS, Georgetown University Law Center, Washington, D.C.
JAMES P. BRUCE, Canadian Climate Program Board, Ottawa, Ontario, Canada
WILLIAM L. FISHER, University of Texas, Austin
EDWARD A. FRIEMAN, Scripps Institution of Oceanography, La Jolla, California
GEORGE M. HORNBERGER, University of Virginia, Charlottesville
W. BARCLAY KAMB, California Institute of Technology, Pasadena
PERRY L. McCARTY, Stanford University, California
S. GEORGE PHILANDER, Princeton University, New Jersey
RAYMOND A. PRICE, Queen's University at Kingston, Ontario, Canada
THOMAS A. SCHELLING, University of Maryland, College Park
ELLEN SILBERGELD, Environmental Defense Fund, Washington, D.C.
STEVEN M. STANLEY, Johns Hopkins University, Baltimore, Maryland
VICTORIA J. TSCHINKEL, Landers and Parsons, Tallahassee, Florida

Staff

STEPHEN RATTIEN, *Executive Director*
STEPHEN D. PARKER, *Associate Executive Director*
MORGAN GOPNIK, *Assistant Executive Director*
JEANETTE SPOON, *Administrative Officer*
SANDI FITZPATRICK, *Administrative Associate*

The National Academy of Sciences is a private, nonprofit, self-perpetuating society of distinguished scholars engaged in scientific and engineering research, dedicated to the furtherance of science and technology and to their use for the general welfare. Upon the authority of the charter granted to it by the Congress in 1863, the Academy has a mandate that requires it to advise the federal government on scientific and technical matters. Dr. Bruce Alberts is president of the National Academy of Sciences.

The National Academy of Engineering was established in 1964, under the charter of the National Academy of Sciences, as a parallel organization of outstanding engineers. It is autonomous in its administration and in the selection of its members, sharing with the National Academy of Sciences the responsibility for advising the federal government. The National Academy of Engineering also sponsors engineering programs aimed at meeting national needs, encourages education and research, and recognizes the superior achievements of engineers. Dr. Robert M. White is president of the National Academy of Engineering.

The Institute of Medicine was established in 1970 by the National Academy of Sciences to secure the services of eminent members of appropriate professions in the examination of policy matters pertaining to the health of the public. The Institute acts under the responsibility given to the National Academy of Sciences by its congressional charter to be an adviser to the federal government and, upon its own initiative, to identify issues of medical care, research, and education. Dr. Kenneth I. Shine is the president of the Institute of Medicine.

The National Research Council was organized by the National Academy of Sciences in 1916 to associate the broad community of science and technology with the Academy's purposes of furthering knowledge and advising the federal government. Functioning in accordance with general policies determined by the Academy, the Council has become the principal operating agency of both the National Academy of Sciences and the National Academy of Engineering in providing services to the government, the public, and the scientific and engineering communities. The Council is administered jointly by both Academies and the Institute of Medicine. Dr. Bruce Alberts and Dr. Robert M. White are chairman and vice chairman, respectively, of the National Research Council.

Foreword

In 1992, the Ocean Studies Board (OSB) of the National Research Council identified seven issues important to society that ocean scientists can and should address. One of these issues is marine biological diversity, as stated in the OSB report *Oceanography in the Next Decade: Building New Partnerships*:

> The ocean comprises a large portion of Earth's biosphere. It hosts a vast diversity of flora and fauna that are critical to Earth's biogeochemical cycles and that serve as an important source of food and pharmaceuticals. In addition to the exciting discoveries of previously unknown biota near hydrothermal vents, many deep-ocean organisms have evolved under relatively stable conditions. Their unique physiologies and biochemistries have not yet been explored adequately, and methods for sampling the more fragile of these species have been developed only in the past decade. Human influence on marine biota has increased dramatically, threatening the stability of coastal ecosystems. Some species have been overharvested; others have been transported inadvertently to areas where they are not indigenous, sometimes resulting in deleterious effects on native species. Still other species are being cultivated commercially, and aquaculture facilities along coastlines are becoming commonplace in some countries. A better understanding of the ecology of marine organisms is urgently needed to prevent irreversible damage to this living resource.[1]

The White House National Science and Technology Council, through its Committee on Environment and Natural Resources (CENR), also identified biodiversity as a critical issue. The CENR appointed a Subcommittee on Biodiver-

---

[1] National Research Council 1992. *Oceanography in the Next Decade: Building New Partnerships*. National Academy Press, Washington, D.C., p. 3.

sity and Ecosystem Dynamics as a mechanism for federal agencies to coordinate biodiversity research.

Because of the importance of marine biological diversity, the OSB and the Board on Biology established a study committee to develop a research strategy to advance our knowledge of factors that affect biological diversity in the ocean. Such a systematic plan is needed to guide research activities toward the common goal of preserving existing diversity in the face of changes brought about by humankind. This report presents the committee's findings and recommendations and includes information and ideas gathered from a broad spectrum of marine scientists.

The research agenda described in this report provides a useful blueprint for U.S. government agencies, the CENR Committee on Biodiversity and Ecosystem Dynamics, the international community, and all those who support and conduct research on biological diversity in marine systems.

William J. Merrell
Chairman, Ocean Studies Board

Preface

Recent widespread changes in the biological diversity of life in the sea are largely due to effects of human activities. Serious concern for the conservation of marine life in the face of rapid population expansion, particularly in coastal regions, and the desire for long-term sustained use of the seas for food, mineral resources, biomedical products, recreation, and other societal needs have motivated marine ecologists and oceanographers to recognize their responsibility to document biodiversity changes and to understand their causes and consequences. The ultimate goal is to improve predictions regarding the magnitude and extent of subsequent alterations to marine biodiversity by human activities. To do this requires a substantially improved understanding of the fundamental mechanisms that create, maintain, and regulate marine biological diversity. Such knowledge is needed to develop conservation and management plans for sustained use by humans of marine habitats and resources that minimize impacts on nature.

Over the last half-decade, the awareness of scientists, society, and state and federal governments regarding crucial issues in biodiversity has been elevated through various grass-roots appeals, such as the *Sustainable Biosphere Initiative* (Lubchenco et al., 1991), *The Diversity of Life* (E.O. Wilson 1992), *The Diversity of Oceanic Life: An Evaluative Overview* (M.N.A. Peterson, 1992), and the *Systematics Agenda 2000* (SA2000, 1994), and through multiagency efforts, such as the *Ecosystem Function of Biological Diversity Programme (Diversitas)* (di Castri and Younès, 1990; Solbrig, 1991), which has a separately identified marine component (J.F. Grassle et al., 1991), and the *Global Marine Biological Diversity Strategy* (Norse, 1993). Most recently, two small, independently organized workshops held in 1993—one sponsored by the National Science Foundation (NSF) (Butman and Carlton, 1993) and one sponsored by the National Research Council's (NRC) Ocean Studies Board (OSB)—spoke in a strong and unified voice of the urgent need for a national marine biological diversity research program.

Responding to cogent arguments and critical needs identified in these and other documents, in November 1993, the OSB and the Board on Biology of the NRC established the Committee on Biological Diversity in Marine Systems to develop the foundation for a national research agenda on this topic based on a workshop of scientists and federal agency representatives. At a meeting held in March 1994, the committee further delineated the purpose of the workshop: to develop a well-defined set of research questions designed to improve understanding of the causes and consequences of changes in marine biological diversity due to effects of human activities, and to provide the knowledge and criteria for assessing and predicting subsequent effects of human stresses on the marine environment.

The workshop, entitled *Biological Diversity in Marine Systems*, was held May 24-26, 1994, at the Arnold and Mabel Beckman Center of the National Academy of Sciences in Irvine, California (Appendix A). Fifty-four individuals participated in the workshop (Appendix B), representing a wide spectrum of expertise in the fields of oceanography, marine ecology, molecular biology, systematics, and population biology. Dynamic discussions among workshop participants resulted in gratifying unanimity on some issues, particularly regarding the perceived most important anthropogenic threats to the marine environment. There was also spirited debate, with invaluable input from the participants, and ultimately a sense of accord and direction was established that crystallized the focus, mission, and substance of this proposed research agenda.

The overall goals, specific objectives, and recommendations for the next steps in developing a marine biodiversity research initiative are discussed herein. The workshop was the primary vehicle by which information was gathered and consensus was built for this report, and the committee deeply appreciates the creative energy and investment of time by the workshop participants. Although the report was written by the committee, many workshop participants contributed both conceptually and in writing, for which the committee is most grateful. We also thank John Ogden, Rita Colwell, Marjorie Reaka, and John Chapman for their useful input. The committee furthermore acknowledges the valuable support of the NRC staff, particularly Mary Hope Katsouros and David Wilmot of the OSB, Eric Fischer of the BB, and Morgan Gopnik of the Commission on Geosciences, Environment, and Resources.

Financial support for the project was provided by the NSF, the Office of Naval Research, the National Oceanic and Atmospheric Administration, the Department of Energy, and the U.S. Department of the Interior's National Biological Service. Representatives from these agencies, particularly Phil Taylor, Randall Alberte, and Michael Sissenwine, contributed to the workshop and provided valuable guidance and insight.

Cheryl Ann Butman
James T. Carlton
Co-Chairs

Contents

xi

Understanding
MARINE
BIODIVERSITY

Executive Summary

The diversity of life in the ocean is being dramatically altered by the rapidly increasing and potentially irreversible effects of activities associated with human population expansion. *Biodiversity* is defined as the collection of genomes, species, and ecosystems occurring in a geographically defined region. The most critical (current or potential) contributors to changes in marine biodiversity are now recognized to be the following: fishing and removal of the ocean's invertebrate and plant stocks, many of which are overexploited; chemical pollution and eutrophication; physical alterations to coastal habitat; invasions of exotic species; and global climate change, including increased ultraviolet radiation and potentially rising temperatures, resulting in possible changes to ocean circulation (and thus nutrient supply and distribution). These stresses to the marine environment have affected and may yet affect life from the intertidal zone to the deep sea.

These activities and phenomena have resulted in clear, serious, and widespread social, economic, and biological impacts including:

- dramatic reductions in most of the preferred edible fish and shellfish species in the world's oceans;
- reduction or loss of species with important potential for biomedical products;
- altered aesthetic and recreational value of many coastal habitats, such as coral reefs, bays, marshes, rocky shores, and beaches;
- vast changes in the species composition and abundance of the ecologically important animals and plants within and between impacted ecosystems; and
- changes in the basic functioning of ecosystems, including the rates and

1

sources of primary production, the stability of populations, the amount and directions of energy flow, and biogeochemical cycling.

Evaluating the scale and consequences of changes in the ocean's biodiversity due to human activities is, however, seriously compromised by critically inadequate knowledge of the patterns and the basic processes that control the diversity of life in the sea.

The basic description of marine biodiversity trails that of the terrestrial realm, particularly as it relates to the extraordinary diversity of higher taxa in the marine compared to the terrestrial environment. Continuing discoveries of new families, orders, and even phyla of marine organisms foretell a wealth of biodiversity yet to be realized.

Like terrestrial habitats, there are vast numbers of undescribed species in familiar oceanic habitats, such as coral reefs and temperate bays and estuaries. There are environments, like the deep sea and polar regions, that are so undersampled that numerous new species are discovered during each expedition to a new area. Newly recognized biological habitats that contain novel species assemblages—such as hydrothermal vents, whale carcasses, brine seeps, and wood debris—continue to emerge, especially in deep water. Moreover, understanding of the mechanisms responsible for the creation, maintenance, and regulation of such habitat-specific marine biodiversity is incomplete, fragmentary, or entirely lacking.

Yet exciting new information, novel techniques, and heightened awareness now permit dramatically improved sampling and species identifications and process-oriented research at increasingly larger geographic scales. Such studies have been previously intractable, but are fundamentally required to understand the consequences of anthropogenic changes to the diversity of marine life.

This report identifies the urgent need for a national research program on Biological Diversity in Marine Systems and outlines a research agenda. *This research agenda proposes a fundamental change in the approach by which biodiversity is measured and studied in the ocean by emphasizing an integrated regional-scale research strategy within an environmentally relevant and socially responsible framework.* This is now possible because of recent technological and conceptual advances within the ecological, molecular, and oceanographic sciences.

Propelled by the need to understand the effects of human activities on biodiversity, this research program would require studies conducted at appropriately large temporal and spatial scales. Given the open nature of marine systems, a regional-scale approach must be taken, one that involves studying multiple, separate sites within a large geographic region. Biological and physical criteria would be used to define this region—that is, to set the maximum spatial and temporal scales required to characterize those processes that control local biodiversity. This decadal-time-scale research program would integrate ecological and oceanographic research spanning a broad range of spatial scales, from local to regional, and over appropriate time scales for distinguishing changes in biodiversity due to effects of human activities from natural phenomena.

A well-defined set of research questions would be addressed in studies of several different kinds of regional-scale marine ecosystems. These studies would permit meaningful comparisons of the causes and consequences of changes in biodiversity due to human activities in different habitats.

This agenda would require significant advances in taxonomic expertise for identifying marine organisms and documenting their distributions, in knowledge of local and regional natural patterns of biodiversity, and in understanding of the processes that create and maintain these patterns in space and time. This would provide, in fact, an exciting opportunity to develop the interface between taxonomy and ecology and between the ecological and oceanographic sciences.

The five fundamental objectives of this first national research agenda on marine biodiversity are:

• to understand the patterns, processes, and consequences of changing marine biological diversity by focusing on critical environmental issues and their threshold effects, and to address these effects at spatial scales from local to regional and at appropriate temporal scales;

• to improve the linkages between the marine ecological and oceanographic sciences by increasing understanding of the connectivity between local, smaller-scale biodiversity patterns and processes and regional, larger-scale oceanographic patterns and processes that may directly impact local phenomena;

• to strengthen and expand the field of marine taxonomy through training, the development of new methodologies, and enhanced information dissemination, and to raise the standard of taxonomic competence in all marine ecological research;

• to facilitate and encourage the incorporation of (1) new technological advances in sampling and sensing instrumentation, experimental techniques, and molecular genetic methods; (2) predictive models for hypothesis development, testing, and extrapolation; and (3) historical perspectives (geological, paleontological, archaeological, and historical records of early explorations) in investigations of the patterns, processes, and consequences of marine biodiversity; and

• to use the new understanding of the patterns, processes, and consequences of marine biodiversity derived from this regional-scale research approach to improve predictions of the impacts of human activities on the marine environment.

As envisioned by the committee, this national research agenda for marine biodiversity would lead to novel, integrated, multiple-scale studies that would improve understanding of how human activities alter marine biodiversity and of why and how such alterations change the functioning of ecosystems. In turn, this understanding would provide valuable information for policymakers regarding the preservation and conservation of marine life, and for identifying those pathways that might save and restore the sea. The ultimate benefit to science and society of this research program would be an enhanced ability for long-term sustained use of the oceans and marine organisms for food, mineral resources, biomedical products, recreation, and other aesthetic and economic gains, while conserving and preserving the diversity and function of life in the sea.

In summary, this marine biodiversity initiative would be:

• An environmentally responsible and socially relevant basic research program on the causes and consequences of changes in marine biological diversity due to effects of human activities.

• A research agenda guided by well-defined research questions that will be addressed concurrently in several different regional-scale systems.

• A program that focuses on large scales that were previously intractable but are absolutely required to address the most compelling biodiversity research questions.

• A partnership between the ecological and oceanographic sciences, both conceptually and methodologically, for explaining biodiversity patterns, processes, and consequences.

• A partnership between ecology and taxonomy, with a major focus on reinvigorating the field of marine taxonomy and systematics.

• A research program with the ultimate goal of improving predictions regarding future effects of human activities on marine biodiversity, thus facilitating the use of the sea for societal needs while minimizing impacts on nature.

Introduction

MARINE BIODIVERSITY IS CHANGING AND IT MATTERS

Approximately 70 percent of the Earth's surface is covered by water, and most of that is marine. Like all biological systems, the oceans are experiencing an ecologically and evolutionarily unprecedented series of stresses, many of which are changing the structure and organization of marine communities. Because humans rely on the oceans for food, mineral resources, and recreation, and because marine life offers potential future benefits to society, such as in the area of biomedical products, it is critical to develop conservation and management strategies that facilitate the long-term sustained use of the sea by humans while minimizing impacts on nature.

Yet to be determined is the ultimate impact of a growing human population on marine biota—from the smallest plankton to the largest whales, living on the bottom or in suspension, at depths ranging from the highest intertidal shores to the abyss (e.g., Boxes 1 and 2). Continuing human population expansion is reflected in alterations to the global atmosphere and heat budgets, changes in hydrologic and sedimentary regimes, chemical contamination and ocean dumping, and the accelerating overexploitation of the ocean's fish and invertebrate stocks (Ray and J.F. Grassle, 1991; M.N.A. Peterson, 1992; Norse, 1993; Weber, 1993; 1994; NRC, 1994a).

Marine life has been altered to large extents, sometimes with dramatic consequences.

• Oyster populations of the Chesapeake Bay that once filtered the entire estuary once a week now filter it only once a year because of stock depletion from overfishing and disease (Newell, 1988).

Box 1: Has human hunting of whales altered deep-sea biodiversity?

WAITING FOR A WHALE:
HUMAN HUNTING AND DEEP-SEA BIODIVERSITY

Organisms in the deep sea are highly food-limited, relying primarily on organic matter raining down from above. Whale carcasses may be particularly important because they are large and sink fast enough with sufficient tissue still intact for exploitation. Lipid-rich whale skeletons have further been found to support an animal community nourished largely by sulfur-reducing chemoautotrophic bacteria. Whale skeleton-associated species are similar, and in some cases identical, to organisms previously thought to be restricted to the chemosynthetic-based hydrothermal vents and other deep-sea microbial reducing habitats. Whale skeletons scattered like islands in the deep sea may thus provide some of the critical stepping stones for organisms between hydrothermal-vent communities, themselves insular and temporary habitats.

Given the potentially important role of whales to deep-sea biodiversity, these communities may have been altered by human hunting of whales. A profound effect of whaling was a vastly reduced—and in some regions, obliterated—whale skeleton supply to the deep sea due to an acute decrease in hunting-generated carcasses (after the turn of this century, whalers retained the entire animal) and to a dramatic decrease in whale populations. This decrease meant a severe spatial interruption, if not elimination, of dispersal corridors between reducing-habitat communities, with potentially marked alterations to biodiversity in hot-vent and cold-seep regions.

Unfortunately, the magnitude and consequences of changes in biodiversity resulting from this type of human activity are difficult to evaluate because of the lack of data on most whale population sizes and distributions, because the region impacted is far removed from the disturbance source, and because effects are being considered nearly a century after the fact. This example does, however, underscore the importance of "thinking big" and "thinking remotely" in evaluating the potential impact of society's activities on nature.

Key References: Stockton and DeLaca (1982); C.R. Smith (1985, 1992); C.R. Smith et al. (1989); Bennett et al. (1994); Deming et al. (in press); Butman et al. (in press).

• The "inexhaustible" fisheries of the great fishing banks (such as Georges Banks and the Grand Banks) are verging on exhaustion or have now been closed (Anthony, 1990), and the historical human hunting of the great whales has resulted in many threatened species (Laws, 1977; Evans, 1987).

• The coral reefs of the Caribbean, Hawaii, and parts of Australasia are threatened by multiple human activities, such as overfishing (Russ, 1991), physical habitat alteration and destruction (Rogers, 1985; Salvat, 1987; WCMC, 1992), and sedimentation associated with logging and land-based development (Hodgson and Dixon, 1988; Kuhlmann, 1988; Ogden, 1988).

Box 2: Did the elimination of large vertebrates such as manatees, turtles, and groupers from tropical ecosystems significantly alter biodiversity?

ELIMINATION OF LARGE VERTEBRATES FROM TROPICAL ECOSYSTEMS

Large marine vertebrates (such as whales, manatees, turtles, groupers, and the extinct Steller's sea cow, Caribbean monk seal, and great auk) have been systematically removed from the oceans by humans over the past 500 years. The ecological effects of the reduction or complete elimination of most large vertebrates remain unknown (sea otter impacts on kelp and sea urchins in the Eastern Pacific are an exception).

Eighteenth-century Caribbean explorers found extraordinarily abundant populations of large vertebrate grazers (manatees, turtles, and parrot fish), large invertebrate grazers (conch snails), and large carnivorous fish (groupers). These animals consumed seagrasses, algae, sea urchins, other fish, and many other animals. The removal of these consumers must have substantially affected communities both directly (e.g., altering food pyramids and trophic structure) and indirectly (e.g., the resulting increases in seagrass populations altering coastal sedimentation processes). Unfortunately, scientists arrived in the Caribbean two centuries after the large animal expulsion commenced.

The elimination of large consumers from a broad region is an example of how early historic alterations at one trophic level can markedly impact modern assumptions and interpretations of both natural biodiversity patterns at other trophic levels and overall ecosystem function. The need for an appropriate retrospective context is clear and further argues for the use of human exclusion experiments (see Box 13) to assess the effects of historical hunting on those species that still survive elsewhere.

Key References: Estes and Palmisano (1974); May et al. (1979); Hay (1984); Thayer et al. (1984); Duggins et al. (1989); Vermeij (1993); Jackson (1994).

• Invasive species are increasing dramatically, with more than 3,000 species a day in motion inside the giant aquaria that serve as ballast tanks in ocean-going vessels (Carlton and Geller, 1993), sometimes completely altering the trophic structure of bays and estuaries into which the ballast water is discharged (e.g., Nichols et al., 1990; Horoshilov, 1993).

• Filling and development of coastal habitat has resulted in total wetland losses of 50 percent in Washington, 74 percent in Maryland and Connecticut, and 91 percent in California (Dahl et al., 1991), with a concomitant loss of critical seagrass and marsh habitat that host a diversity of invertebrates and fish (including many economically important species), protect coastlines from erosion, enhance nutrient cycling, and improve water clarity (Fenchel, 1977).

There are thus significant reasons for concern. The timing is critical for determining the processes that contribute to these fundamental changes and for developing a predictive understanding that will allow preservation and restoration of the ocean's biodiversity. The dual issues of change and loss of marine biodiversity are not trivial and have unified marine scientists—oceanographers, ecologists, and taxonomists—in a common cause.

National and international social and economic implications of accelerating change bear directly on interrelated subjects, such as:

- the ocean's capacity to sustain economically significant fisheries,
- the quality of bays and estuaries as nurseries for important stocks,
- the loss of species with important potential for biomedical products,
- the increasingly chronic nature of blooms of toxic algae,
- the recreational value of ocean margins, and
- the aesthetic value of marine environments that remain close to their aboriginal state.

Marine biological diversity is changing, and *it does matter*.

This document identifies the urgent need for a national research program on the biological diversity of marine systems. In this research plan, *biodiversity* is defined as the collection of genomes, species, and ecosystems occurring in a geographically defined region. This agenda focuses on a novel program where ecological and oceanographic research would be integrated at all relevant spatial scales, from local to regional, and over appropriate time scales for distinguishing changes in biodiversity due to effects of human activities from natural phenomena. Integral to this initiative are *taxonomy* (here defined as the descriptive branch of the larger field of systematics) for documenting the magnitude and patterns of biodiversity, and predictive *models* for hypothesis development, testing, and extrapolation, and for developing guidelines for management and conservation.

THE DEPTH AND BREADTH OF
UNDERDESCRIBED MARINE BIODIVERSITY

"The future historians of science may well find that a crisis that was upon us at the end of the 20th century was the extinction of the systematist, the extinction of the naturalist, the extinction of the biogeographer—those who would tell the tales of the potential demise of global marine diversity."

Carlton (1993, pg. 507)

There are many exciting recent discoveries of previously unknown marine organisms that form critical links in ecosystem function. These discoveries often were facilitated by the development of new sampling and analytical techniques and emphasize a science that has been exploring just the periphery of the biodiversity frontier in the oceans. Examples include the following:

• The brown tides that led to the demise of the bay scallop industry of southern New England in the 1980s were caused by a protist that was previously unknown and had no genus or species name (Sieburth et al., 1988).

• Major estuarine fish kills have now been associated with a previously undescribed "phantom" dinoflagellate whose existence and identity were only announced in 1992 (Burkholder et al., 1992).

• In the open ocean, the prochlorophytes, a group of marine, free-living, bacterial primary producers, were not discovered until the late 1980s—and yet they are now known to account for up to 40 percent of the chlorophyll in some ocean regions (Chisholm et al., 1988; R.J. Olson et al., 1990; S.W. Chisholm, pers. comm., 1994).

• Species of marine nonphotosynthetic microbes—bacterial taxa now known as Eubacteria and Archaea—are now being discovered at a rapid rate thanks to new molecular genetic techniques and are the basis (along with enhanced awareness of the widespread abundances of marine viruses [Bergh et al., 1989; Proctor and Fuhrman, 1990]) for rapidly evolving concepts of marine microbial diversity and the role of microbes in global geochemical cycles (Box 3).

• What was once thought to be a single species of algal symbiont in the Caribbean star coral (that is in fact at least three coral species, as discussed later) is now known to be three major groups of symbionts (Rowan and Powers, 1991, 1992) that occur at different depth zones (Rowan and Knowlton, in press). The ecological significance of these findings is currently being investigated, including the possibility that the symbiont species have different propensities for expulsion from the three coral species, which may help to explain the phenomenon of variable coral-bleaching episodes (N. Knowlton, pers. comm., 1994).

There are numerous undescribed species in even the most familiar of ocean environments—ranging from the common harpacticoid copepods and worms of shelf muds to the tiny nematodes and highly colorful sea slugs in tropical lagoons. Table 1 provides a glimpse of the magnitude of underdescribed diversity of marine invertebrates at the level of individual taxa. At the level of entire ecosystems, there are environments such as the deep sea and polar regions that are so undersampled that hundreds of new species are discovered during each expedition to a new area. Indeed, knowledge of the ecology and evolutionary history of the deep sea has been fundamentally altered by the discovery that the diversity of abyssal communities is dramatically higher than previously thought (J.F. Grassle, 1991; J.F. Grassle and Maciolek, 1992).

None of these estimates of underdescribed marine biodiversity takes into account the innumerable and ecologically important benthic and planktonic protists, which alone may comprise at least 34 phyla and 83 classes (Corliss, 1994), nor the vast complexity of the undescribed parasites that live on and in other marine organisms. New views of the true scale of marine biodiversity dictate substantial rethinking of current understanding of the processes that create and

Box 3: Knowledge of marine microbial biodiversity is being revolutionized by the application of molecular genetic techniques.

MOLECULAR ECOLOGY AND SYSTEMATICS: PROVIDING NEW PERSPECTIVES ON MARINE MICROBIAL DIVERSITY

The application of molecular genetic techniques and approaches is now providing a remarkable new perspective on the biodiversity of the abundant and ubiquitous planktonic bacterial assemblages in the oceans. Oceanic microbes are integral components of marine food webs: they often dominate the plankton biomass, they are extremely important in the Earth's biogeochemical cycles, and they are responsible for much of the cycling of organic matter in the sea. The biodiversity of these assemblages is virtually undescribed.

Conventional identification methods involving culturing have identified an estimated 1 percent of the microbes present in marine samples. With molecular techniques it is now possible to begin to determine the composition, diversity, and variability of the remaining 99 percent of the marine bacterioplankton. The vast prokaryotic assemblages found in the food-limited (oligotrophic) open ocean, for example, are now being described, with important implications for understanding their functions in these ecosystems and their responses to environmental perturbation.

The recent discovery of abundant and widespread new *phyla* of microorganisms underscores the extent of ignorance with regard to microbial biodiversity. Indeed, in some cases these organisms are known solely from their DNA sequences. The application of molecular techniques to discover new microbial groups in the ocean, supplementing traditional microbiological techniques, throws open a door to the unknown: for example, fundamentally new types of organisms appeared in only the first few plankton samples examined with molecular genetic techniques from one small site in the open ocean off of San Diego, California.

Key References: DeLong et al. (1989, 1993); DeLong (1992); Fuhrman et al. (1992, 1993); Giovannoni et al. (1990, 1993); Schmidt et al. (1991); Ward et al. (1990).

maintain diversity, and of the impacts of human society on ocean ecosystems. It is probable that these breakthroughs represent only the smallest fraction of the exciting discoveries that lie ahead. *Most* of the species in the ocean may still remain undiscovered and undescribed (e.g., Table 1).

These gaps in knowledge of the magnitude and extent of marine biodiversity have real consequences. Inadequate knowledge of the species present in a given marine community or ecosystem limits understanding of ecosystem function and predictions of how human alterations impact that function. Understanding which species are critical in energy flow from lower to higher trophic levels in a food chain, for example, may be nearly impossible if many members of a particular group of prey or predators are undetected or undescribed. As noted elsewhere in

TABLE 1 Examples of the Magnitude of Underdescribed Biodiversity Among Marine Invertebrates in Familiar and Easily Accessible Marine Environments

Site	Taxon	Number of Undescribed Species out of Total Collected in the Taxon	Source[a]
Gulf of Mexico	Copepods (harpacticoids)	19-27 of 29 (shelf site, 18 m)	D. Thistle
New Guinea	Snails, sea slugs	310 of 564 (one lagoon)	T. Gosliner
Philippines	Snails, sea slugs	135 of 320 (one island, multiple sites)	T. Gosliner
Georges Bank	Marine polychaete worms	124 of 372 (shallow shelf, multiple stations)	J. Blake
Hawaii	Marine polychaete worms	112 of 158 (6 liters of coral reef sediment, one island)	Dutch (1988)
Great Barrier Reef	Marine flatworms (polyclads)	123 of 134 (two islands)	Newman and Cannon (1994)

[a]Personal communication to J.T. Carlton unless indicated otherwise.

NOTE: The third column represents an estimate of the number of undescribed species of a given taxon found in samples taken at the site and habitat indicated. Sampling effort and number of samples varied among studies.

this report, this does not mean that every species in a system must be described in order to understand that system. Rather, *sufficient* knowledge of the breadth and depth of the diversity of animals, plants, microbes, and other life present at a site or in a region is needed to understand the ecological roles of abundant and critical species and the functioning of the ecosystem. These considerations are not limited to ecological interest; as with the discovery of previously newly described terrestrial species which proved to be of biomedical value, the next newly described organism in the sea could prove to be a key species in the rapidly developing field of marine biotechnology (see Box 14).

SIGNIFICANT OPPORTUNITIES FOR FORGING NEW HORIZONS

Although the number of undescribed, underdescribed, and inaccurately described species in the oceans appears daunting, new techniques and approaches are rapidly improving the ability to detect and describe the genetic, species, and ecosystem diversity of the oceans.

Genetic Diversity

With new molecular techniques, surprising levels of genetic diversity are now being discovered in marine organisms ranging from phytoplankton (Wood and Leathem, 1992) to sea turtles (Bowen et al., 1991, 1993), often calling into question critical concepts of speciation in the sea (Palumbi, 1992). Application of molecular techniques for evaluating intraspecific genetic diversity has further provided valuable information for the identification and management of endangered species. Evidence from mitochondrial DNA (mtDNA) analysis has shown, for example, that the last remaining and highly restricted population of several hundred Kemp's ridley sea turtles is, in fact, genetically distinct from the closely related sister species, the olive ridley, that is more widespread, thus confirming the need to protect the former species (Bowen et al., 1991). Recent mtDNA analysis of humpback whales has shown genetic differences over surprisingly short distances with important implications for conservation (see Box 11).

Species Diversity

Molecular genetic techniques combined with classic morphometric approaches are now revealing numerous sibling species complexes within what were frequently believed to be single species (Knowlton, 1993). Examples include the following:

• Perhaps one of the world's best-known marine invertebrates, the mussel *Mytilus edulis*, is now known to be three distinct species (McDonald et al., 1992)—and yet this mussel has formed the basis, on the presumption that it was *one* species, for the pollution-monitoring "International Mussel Watch Program" (NRC, 1980). The different growth rates of at least two of these cryptic species evidently result in observed different body burdens of some contaminants (Lobel et al., 1990), leading to further explorations of the implications of species-specific variations in contaminant uptake on comparative programs like Mussel Watch (B. Tripp, pers. comm., 1994).

• A number of abundant and widespread tropical species of corals and bryozoans, the subjects of intensive ecological studies, are now known to be species complexes (e.g., Knowlton and Jackson, 1994). A particularly striking example is the common Caribbean star coral *Montastraea annularis*, which is now known to be at least three distinct species (Knowlton et al., 1992). Furthermore, the three species have different growth rates and carbon isotope ratios, parameters routinely used to estimate past climatic conditions, thus affecting previous estimates of global climate change.

• The marine worm *Capitella "capitata"* was once regarded as a cosmopolitan "indicator" species of disturbed, organic-enriched sediments, but it is now known to be 15 or more sibling species that occur from the intertidal zone to the deep sea (J.P. Grassle and J.F. Grassle, 1976; J.F. Grassle and J.P. Grassle, 1978;

J.P. Grassle, 1980; J.P. Grassle, pers. comm., 1993). Most of the species differ in some of their life-history characteristics, such as larval development type, brood size, and generation time. Thus, the use of sibling species within this complex as bioassays of environmental degradation hinges on understanding the ecological consequences of their life-history variation.

• Cryptic sibling species have now been discovered in important commercial species, including the oyster *Crassostrea*, the shrimp *Penaeus*, and the stone crab *Menippe*, with important implications for conservation and management (Knowlton, 1993). Examples exist for both exploited and protected species. Identification of the Spanish mackerel *Scomberomorus maculatus* as two species that mature at different ages and sizes (Collette et al., 1978) dictates the avoidance of using life history data of the first species for management of the second species. Use of molecular techniques has also suggested that the common dolphin (*Delphinus delphis*) is actually two species that may have different distributions and abundances (Rosel et al., 1994), and therefore different requirements for protection.

These types of discoveries help to foster an appreciation of the true extent of marine biodiversity, and add a considerable new dimension to estimates of how many species exist in the ocean. It is clear, however, that molecular techniques provide one of the most powerful means for revealing a new understanding of the ocean's complexity (see Box 3).

Habitat and Ecosystem Diversity

Advanced instrumentation and sampling have revealed new species assemblages in novel habitats in the oceans, such as hydrothermal vents (J.F. Grassle, 1986; Tunnicliffe, 1991), whale carcasses (C.R. Smith et al., 1989), wood debris (Turner, 1973, 1981), and sites of hydrothermal, brine, and hydrocarbon seepage (Williams, 1988; Kennicutt et al., 1989; Southward, 1989; MacDonald et al., 1990). As with the discovery of new genomes and new species, it is doubtful that hydrothermal vents or whale skeleton biotas have closed the final chapter on the discovery of novel habitats or ecosystems in the sea. Are there, for example, unique biotas in the abyssal depths of the mid-Atlantic Ocean singularly tied to sinking masses of the pelagic seaweed *Sargassum*? Moreover, new discoveries are unlikely to be limited to just the vast deep-sea depths. Discovery of novel chemoautotrophic associations, shallow and deep (reviewed in Bennett et al., 1994), is a striking reminder that the biodiversity of the majority of the Earth's surface may be dependent on yet undiscovered and unanticipated habitat diversity. Continually improving capabilities for exploring large regions of the ocean floor and water column (e.g., see Box 12) now set the stage for searching the sea in ways impossible to imagine only a few years ago.

Linking Pattern to Process:
A Regional-Scale Approach

A research program to understand the role of human activities in altering biodiversity must involve studies at relatively large spatial and temporal scales. The open nature of marine systems (Box 4) dictates a *regional-scale approach* that involves studying *multiple separate sites within an appropriately large geographic region.* Physical (together with chemical and geological) oceanographic processes provide the dynamic backdrop against which all biological processes take place in the ocean. Thus, biological and physical considerations should be used to delimit regional-scale systems. Water motion affects biology by acting as a transport mechanism for organisms and their propagules, as a dynamic boundary between regimes, and as a force to which organisms must adapt or respond, for example, in their feeding and locomotor activities (Vogel, 1981; Nowell and Jumars, 1984; Denny, 1988, 1993).

Delineating the boundaries of geographic regions thus involves both biological and physical criteria—criteria that define the maximum spatial and temporal scales generally required to characterize adequately the processes that control biodiversity maintenance at a *local* scale. This means, for example (as explained below) that it is imprudent to study long-term prospects for biodiversity in the Chesapeake Bay independently of other Atlantic coastal plain estuaries, or to assume that the coral reefs of Florida are independent of other reef tracts in the greater Caribbean area (see Box 5).

Fundamental research questions concerning the creation, maintenance, and regulation of biodiversity have generally been studied experimentally on very small scales (centimeters to hundreds of meters). Consequently, much is known about mechanisms of interaction and the proximate causes of local dynamics, much less about how the sum total of species at a site indirectly affects individual

Box 4: Marine ecosystems have many attributes distinct from those of terrestrial ecosystems that have important implications for understanding biodiversity.

DISTINCTIVE FEATURES OF MARINE ECOSYSTEMS

Marine and terrestrial ecosystems differ in significant ways that suggest that the ocean may respond to human perturbations in a fundamentally different manner from the land. Some of the unique attributes of marine ecosystems are listed below:

• Marine primary producers are represented by small and often mobile phyla. Terrestrial producers tend to be large and sessile. Marine producers are subject to fluid transport processes, can be spatially mixed, and can unexpectedly produce blooms that may be toxic.

• Large marine carnivores and grazers—top predators such as fish and sea stars—have a greater range of life-history characteristics than terrestrial counterparts. Most marine predators have planktonic and benthic life stages, each with unique environmental responses. Marine predators differ strikingly in their much higher reproductive output. This may buffer them from extinction due to overexploitation, but it also renders their populations far more variable and less predictable and makes them more vulnerable to threshold effects.

• When ocean and continental (aquatic and terrestrial) systems are compared, biomass is found to be thousands, to hundreds of thousands of times more dilute in the oceans, oceanic species interact trophically with more other species than continental species, the largest marine predators and prey are larger by one or two orders of magnitude, and the oceans are on average several to hundreds of times less productive than the continents.

• Distant marine habitats can be linked by dispersing larvae. Such systems are "open," and connections between benthic and planktonic life-history stages assume great significance, unlike most terrestrial systems.

• The higher order diversity of marine life is substantially richer: there are 13 unique marine animal phyla (as opposed to 1 unique land phylum). The existence of such a large number of unique phyla provides a compelling argument for the importance of the evolutionary history of life in the sea.

Key References: Steele (1985); May (1988); Steele et al. (1989); Strathmann (1990); Cohen (1994); Knowlton (1993).

species, and almost nothing about interactions among communities at different sites (e.g., S.A. Levin, 1992). Changes in climate and the flux of nutrients, plankton, or larval recruits all depend on regional oceanographic processes, often largely independent of local events (e.g., Butman, 1987; Roughgarden et al., 1988; Underwood and Petraitis, 1993). There are also nagging questions about minimum population sizes and critical areas for survival of species in the sea (e.g., Nee and May, 1992). None of these issues can be addressed at a local scale.

Moreover, understanding larger-scale phenomena is central to any scientifically rational plan for biodiversity conservation and maintenance.

THE SPATIAL CONTINUUM

". . . the problem is not to choose the correct scale of description, but rather to recognize that change is taking place on many scales at the same time, and that it is the interaction among phenomena on different scales that must occupy our attention."

S.A. Levin (1992, p. 1947)

Biodiversity depends on processes operating at many different spatial, temporal, and organizational scales (S.A. Levin, 1992). In general, these scales are broadly overlapping, with processes interacting among scales (Fig. 1). The most relevant scales for studying particular species or dynamical interactions among species will not be the same as for others, and hence there is no single correct set of scales for viewing a system. Rather, there must be awareness of the selective filter a particular perspective imposes on observed dynamics and of how information is transferred across scales (Ricklefs, 1987, 1990; Underwood and Petraitis, 1993).

A "patch" is an initially relatively uniform portion of a habitat whose limits

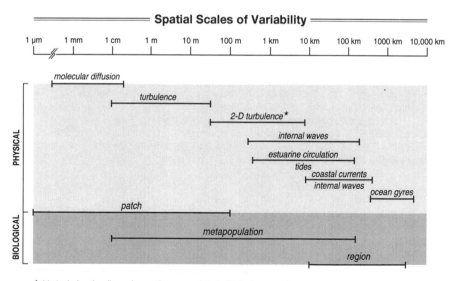

===== **Spatial Scales of Variability** =====

| 1 μm | 1 mm | 1 cm | 1 m | 10 m | 100 m | 1 km | 10 km | 100 km | 1000 km | 10,000 km |

molecular diffusion

turbulence

2-D turbulence*

internal waves

estuarine circulation

tides

coastal currents

internal waves

ocean gyres

PHYSICAL

patch

BIOLOGICAL

metapopulation

region

* this includes the dispersive motions associated with features such as island wakes, headland eddies, instabilities and Gulf Stream rings

FIGURE 1 The spatial scales of variability associated with physical and biological processes (adapted from NRC, 1987).

are defined by relatively abrupt changes in the abundance of common species (Paine and S.A. Levin, 1981), or by the physical phenomena contributing to patch formation and maintenance (Fig. 1). This is the scale over which individual organisms interact with each other and with their immediate environment. Patches in the pelagic environment are three-dimensional and mobile, whether they consist of actively moving nekton (Hamner et al., 1983; Boudreau, 1992) or passively advected plankton (Haury et al., 1978; Fasham, 1978; Yoder et al., 1981). In pelagic systems, patches are found on almost every scale of observation (Owen, 1989; Powell and Okubo, 1994), and measures of patchiness seem to change continuously across scales (e.g., S.A. Levin et al., 1989). Defining and understanding characteristic patch sizes for plankton thus remain a focus of considerable interest, and although patchiness in the pelagic environment is less well-understood than in some benthic systems, new acoustic and imaging technologies are increasing the ability to measure patchiness (see Box 12). For intertidal and sublittoral benthic communities, distinct patches are much more evident and may impose a hierarchical organization to the system (Paine and S.A. Levin, 1981). In these systems, interindividual and interspecific interactions occur to a larger extent within patches, and processes such as dispersal, disturbance, and predation impose interpatch correlations (Sousa, 1985).

The collection of patches that interact or are otherwise connected constitute a "metapopulation" (Gilpin and Hanski, 1991). Note that this includes empty patches of suitable habitat, for example, in a mussel bed or on a coral reef, where a group of mussels or corals has been recently consumed by sea stars or destroyed by moving debris during violent storms (Connell, 1978; Paine and S.A. Levin, 1981). The subsequent successional events within each patch, and the movements of organisms among patches, can be integrated to characterize the demography of patches at the site. Physical processes operating over a wide range of time and space scales may affect metapopulation dynamics (Fig. 1). The theory of metapopulations is still in its infancy, and few marine populations have been studied at this scale. Communities are the sum total of metapopulations in a given habitat at a site.

A "region" comprises all the sites within a biogeographic province whose limits are defined by the relative homogeneity of the biota, as well as by unifying geographic features and oceanographic processes (Fig. 2). Ecological connections between sites within a region depend on fundamentally different processes from those between patches within sites (Butman, 1987; Roughgarden et al., 1988; Underwood and Petraitis, 1993). For example, the exchange of nutrients, food, and colonists between one site and another all depend on patterns of coastal and ocean currents that are largely independent of local events (e.g., see Box 5). Such processes can be modeled as a metapopulation of sites or metacommunities (Gilpin and Hanski, 1991; D.S. Wilson, 1992). There are, however, almost no data showing how changes in biodiversity at one site within a region affect the diversity at another.

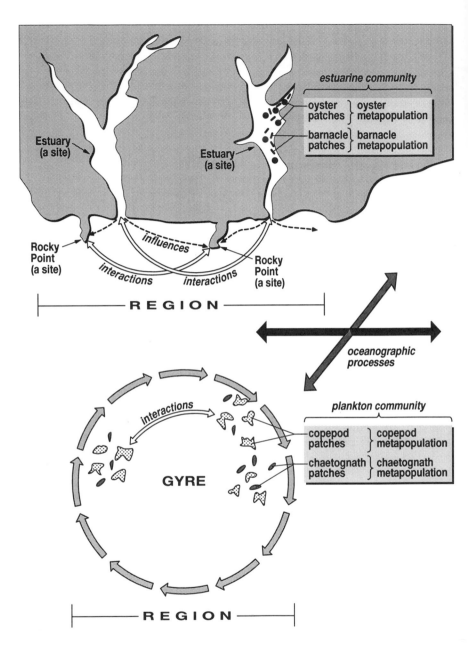

FIGURE 2   The relationship between patches, metapopulations, communities, sites, and regions using a benthic and a pelagic example.

Examples of coastal regional-scale systems include the *region* that contains all the coral reefs of the Hawaiian Islands, the *region* that contains all the rocky open coasts of the Pacific Northwest, or the *region* that contains all the estuaries and enclosed bays of the Gulf of Mexico coastal plain. These regions thus encompass all physical and biological processes that could affect the reefs, rocky shores, or estuaries and bays contained within them. For example, the frequency, magnitude, and spatial extent of freshwater outflow from major rivers or local estuaries may affect coastal communities, and, likewise, offshore circulation is one factor controlling the residence time of water (containing larvae and nutrients) within bays and estuaries.

Within pelagic open-ocean systems a region could be specifically defined by a physical-oceanographic feature, by relatively persistent circulation patterns, or by water-property distributions. Thus, physical-oceanographic processes that occur on a variety of spatial scales (see Fig. 1) are critical elements to the definition of a pelagic regional-scale system (Fig. 2). Present-day oceanographic features that are persistent over relatively long time scales (days to months or longer) are likely to have had the most impact on defining, from a physical perspective, contemporary biological boundaries (but see below regarding longer time scales). Well-known examples of persistent physical features in North America include tidal mixing fronts (e.g., Gulf of Maine), gyres (e.g., North Pacific, Bering Sea), upwelling fronts (e.g., coastal Eastern Pacific), and warm-core rings (e.g., North Atlantic).

TEMPORAL CONSIDERATIONS

"Geological history and oceanographic processes are the warp and woof of the biological understanding of any marine habitat."

Dayton et al. (1994, p. 90)

Biological patterns and processes also vary over a range of temporal scales. Time scales associated with speciation may range from thousands to millions of years, depending on such factors as the mode of speciation and the ecology and life-history characteristics of the organisms involved. Yet, time scales for genetic changes within species may be much less—on the order of years to centuries. Almost nothing is known about appropriate time scales for significant genetic changes in microorganisms. But, clearly, ecological interactions among members of a community, and adaptations of organisms to their environment, may develop over very long time scales indeed. In fact, because current regimes and other physical transport characteristics of the marine environment have changed over geologic time due to changing sea level, coastline, and bottom morphology, and water temperature and salinity characteristics, it cannot be assumed that organisms are necessarily optimally adapted to the physical regime in which they presently reside (Valentine and Jablonski, 1993). Any understanding of present-day patterns of biodiversity therefore must be embedded firmly in an historical perspective, underscoring the importance, for example, of knowing if a given species was introduced or is native to the environment in which it now occurs (Carlton, 1989; Valentine and Jablonski, 1993) and of retrospective analyses in general (discussed in Chapter 6).

Understanding and predicting changes in biodiversity due to effects of human activities requires consideration of the time scales of variation of physical processes relative to biological processes. This consideration defines the relevant regional scale of study for a given community or habitat. This marine biodiversity initiative is envisioned as a decadal-scale program. Given this time frame and physical considerations, important life-history features of the organisms include, for example, generation time, larval type (direct development or planktonic), dispersal period in the plankton (for pelagic larvae or spores of benthic organisms), and resting stage duration (for dormant benthic stages of planktonic organisms). Generation time, for example, is one criterion dictating the ultimate *duration* of a study whose goal is to understand *changes* in biodiversity due to effects of human activities.

Moreover, spatial scales tend to increase as time scales increase. Thus, for example, time scales associated with various physical oceanographic processes define advection length scales for planktonic larval transport, and the frequency and duration of resuspension events for benthic resting stages. Based on larval transport considerations alone, then, regional spatial scales would tend to be smaller for polar compared with temperate benthic invertebrate communities because the proportion of direct developers tends to increase with latitude (Thorson, 1950; Mileikovsky, 1971; Strathmann, 1985).

DEFINING THE MOST MEANINGFUL SCALES OF STUDY

Ultimately, a variety of criteria, based on all life-history stages of the organisms, the oceanographic processes relevant to these stages, and the critical environmental issues in the region, must be used in defining the maximum spatial and temporal scales of study (e.g., Box 5). Once the maxima have been defined, shrewd insight will be needed to select the most meaningful smaller scales of study. A compelling terrestrial example is the 12-year study of Tilman and Downing (1994; but see also Givnish, 1994 and Tilman et al., 1994) on the relationship between the biodiversity of vascular plants (manipulated by nitrogen fertilization) and the stability of the grassland ecosystem (in response to drought). This marine biodiversity program would emphasize inclusion of biological and oceanographic processes at the largest relevant scales—recognizing the connectivity among sites within a geographic region—while acknowledging the critical importance of ecological processes and physical-biological coupling at smaller, site-specific scales. This initiative does *not* recommend studying all processes at all scales within a selected geographic region. Rather, this initiative

---

Box 5: Understanding the origins of profound biological changes at one site will require understanding similar or related changes at the regional level.

FLORIDA BAY: THE NEED TO SEEK A BIGGER PICTURE

In what may be a system that mirrors the broad range of human alterations to estuaries, there is disturbing evidence that the Florida Bay ecosystem is collapsing. Florida Bay (2,200 km^2), and the adjacent Florida Keys, is the only tropical marine ecosystem in the continental United States, with a vast economic value for the state of Florida. Florida Bay and the coral reef tracts of the Keys are connected by coastal currents, and thus the Bay may have a critical influence on the reefs. Likewise, alterations to land runoff and freshwater systems can affect Bay waters.

• In 1987, seagrasses began to die and this has affected over 18 percent of the Bay as of 1993.
• Because algal blooms have increased, large areas of the Bay are now subjected to bloom conditions, and these have spread to the coral reefs of the Keys.
• Populations of water birds, forage fish, and juvenile gamefish are significantly reduced in eastern Florida Bay where freshwater inflow from the Everglades has been reduced and hypersalinity results.
• Catches in the Dry Tortugas of commercially valuable pink shrimp, which spend their juvenile stages in Florida Bay, have declined considerably. Large sponges, important for spiny lobster habitat, have also declined.

There is considerable scientific debate over the causes of collapse, and thus of the potential for recovery, or the steps that are necessary to implement and facilitate recovery. Whereas some of these changes could be considered as only "local" effects, it is important to view this ecosystem collapse in terms of broader changes in the Caribbean marine and terrestrial ecosystems. A more narrow focus on only Florida Bay may be doomed to failure. An important further lesson of the Florida Bay situation is that the water systems on land that impact the Bay, and in turn the Keys offshore, also must be understood and managed.

Key References: Robblee et al. (1991); Porter and Meier (1992); Boesch et al. (1993); Ogden et al. (1994).

---

is designed to accommodate studies at all scales *relevant* to a specified biodiversity research project.

A RATIONALE FOR STUDYING SEVERAL
TYPES OF REGIONAL MODEL SYSTEMS

Three biological rationales suggest the need to use a regional-scale approach to study, concurrently, a variety of different *types* of marine ecosystems.

• Various marine systems present different opportunities to address fundamental questions because the systems vary in their evolutionary history, resulting
in different biotas; in their connectivity (among sites within a system and among
systems); and in the knowledge and description of their biodiversity.

• Establishing the appropriate regional scale is simpler for systems whose
boundaries are more easily defined, and for which there is greater knowledge of
specific anthropogenic stresses and their effects, as well as greater understanding
of the role of dispersal mechanisms.

• There are fundamentally different levels of environmental impact in different marine systems: some are now highly impacted (such as most coastal and
coral-reef systems), whereas for others there is a grave possibility of future impact (such as open-ocean and deep-sea systems through, for example, human
alterations to the global climate) suggesting the advisability of "anticipatory"
research.

At larger—possibly the largest—scales, it is clear that distinct regional systems thousands of kilometers apart may have reciprocal influences on each other.
Currents carry larvae of some species transoceanic distances, far from their par-
ents (Scheltema, 1986). For example, the larvae of the Caribbean spiny lobster
spend 9 or more months traversing the open-ocean pelagic ecosystems of the
North Atlantic gyre (Farmer et al., 1989). Thus, an important factor influencing
the abundance of adult lobster populations may be very distant oceanographic
and biological processes—in an ecosystem not likely to be directly studied by
Caribbean lobster ecologists. Likewise, larvae of the American eel undergo
planktonic development in the Sargasso Sea, thousands of kilometers from the
adult habitat in coastal rivers (Avise et al., 1986; McCleave, 1993). Also, currents associated with periodic El Niño-Southern Oscillations are known to transport warm water and propagules across the Pacific, causing population and community changes (Glynn, 1988).

Although this marine biodiversity initiative focuses on the larger within-
regional-scale oceanographic and ecological processes that directly bear on the
expression of local, site-specific biodiversity, awareness of the ocean-wide influences *between regions* remains important to an overall understanding of regional
biodiversity. Furthermore, addressing a well-defined set of research questions in
regional-scale studies of several different types of marine ecosystems will permit
meaningful comparisons of the causes and consequences of changes in biodiversity due to human activities. Such comparisons will greatly enhance predictive capability relative to future human impacts on the marine environment.

OBJECTIVES OF THIS INITIATIVE

This marine biodiversity initiative sets a research agenda that represents a
fundamental change in the approach to measuring and studying biodiversity in

the oceans by emphasizing an *integrated regional-scale research strategy* within an environmentally relevant and socially responsible framework. The purposes of this proposed initiative are thus to define a suite of cross-system basic research questions that transcend the typical approach to studying the patterns, processes, and consequences of marine biological diversity. These questions are designed to yield new insights and to enhance understanding of the factors controlling biological diversity, and to increase predictive capability regarding changes in marine biodiversity due to anthropogenic effects in order to provide essential management and conservation guidelines.

The five fundamental objectives of this initiative are:

• to understand the patterns, processes, and consequences of changing marine biological diversity by focusing on critical environmental issues and their threshold effects, and to address these effects at spatial scales from local to regional and at appropriate temporal scales;

• to improve the linkages between the marine ecological and oceanographic sciences by increasing understanding of the connectivity between local, smaller-scale biodiversity patterns and processes, and regional, larger-scale oceanographic patterns and processes that may directly impact local phenomena;

• to strengthen and expand the field of marine taxonomy through training, the development of new methodologies, and enhanced information dissemination, and to raise the standard of taxonomic competence in all marine ecological research;

• to facilitate and encourage the incorporation of (1) new technological advances in sampling and sensing instrumentation, experimental techniques, and molecular genetic methods, (2) predictive models for hypothesis development, testing, and extrapolation, and (3) historical perspectives (geological, paleontological, archeological, and historical records of early explorations) in investigations of the patterns, processes, and consequences of marine biodiversity; and

• to use the new understanding of the patterns, processes, and consequences of marine biodiversity derived from this regional-scale research approach to improve predictions of the impacts of human activities on the marine environment.

THREE

Biodiversity Changes Due To Anthropogenic Effects: Critical Environmental Issues

"The ocean comprises a large portion of Earth's biosphere [and] hosts a vast diversity of flora and fauna that are critical to Earth's biogeochemical cycles and serve as an important source of food and pharmaceuticals. . . . Human influence on marine biota has increased dramatically, threatening the stability of coastal ecosystems."

National Research Council (1992, p. 3)

Human activities, directly and indirectly, are now the primary cause of changes to marine biodiversity. Natural perturbations have always occurred in the oceans—ranging from seasonal climatic events (such as hurricanes, typhoons, and storm tides) leading to local habitat destruction, to El Niño-Southern Oscillation events, to natural oil seeps—but the resulting changes in biodiversity were frequently reversible or have been long integrated into the larger spatial and temporal patterns of ecosystem structure and function. Effects of many human activities, however, are frequently irreversible, at least over the span of a human life.

Human activities that affect biodiversity are here referred to as *critical environmental issues*. These issues provide the focus for prioritizing research on the patterns, processes, and consequences of marine biodiversity. In turn, intimate knowledge of how human perturbations affect biodiversity ultimately provides clearer insight regarding the symptoms of changes in the sea caused by human activities.

A vast literature has addressed the types, causes, and significance of anthropogenic stresses in the oceans (e.g., Kinne, 1984; Sherman et al., 1990; Beatley, 1991; GESAMP, 1991; Clark, 1992; Thayer, 1992; Norse, 1993; C.H. Peterson, 1993; Suchanek, 1994). Significant criteria in recognizing those stresses most important to changes in biodiversity include (1) their ubiquity across many different marine habitats, (2) their duration and magnitude, and (3) their degree of reversibility. Moreover, the strength and impact of many human perturbations will vary because of interactions with natural environmental conditions and changes. In turn, alterations to biodiversity may result from (1) the cumulative effects of one or more stresses, (2) the synergistic effects of two or more stresses, or (3) the cumulative or synergistic interactions between natural and human stresses.

Given the above criteria for identifying critical environmental issues, the committee believes the most important agents of present and potential change to marine biodiversity at the genetic, species, and ecosystem levels are the following five activities:

- fisheries operations,
- chemical pollution and eutrophication,
- alteration of physical habitat,
- invasions of exotic species, and
- global climate change.

Singly or in combination, these human perturbations can lead to transformation of energy flow patterns and many other fundamental alterations in system structure and function (e.g., Boxes 5 and 6). Human activities have further led to the global extinction of marine mammals, birds (Vermeij, 1993), and invertebrates (Carlton, 1993), although little is known about the number of threatened, endangered, or extinct marine invertebrates or fish (e.g., Lovejoy, 1980). Many species of marine animals have been hunted to commercial and ecological extinction (Norse, 1993), with potential genetic consequences and concomitant implications for management and conservation programs (Ehrlich and Ehrlich, 1981). Particularly difficult management issues and questions emerge for species that are sliding along the continuum from threatened to endangered.

There is no one operational definition of "serious change" in the oceans—rather, the seriousness of a change is a function of the balance between the magnitude and persistence (endurance) of a perturbation on the one hand, and the ability of a given system to recover from that disturbance when its effects are reduced or stopped on the other. It is evident that a thorough understanding of the composition and functioning of an ecosystem is fundamental to recognizing changes in that system. However, it will not always be possible to separate long-term natural variation or cyclic changes from human impacts and the potentially large synergistic interactions between them. The ubiquity and magnitude of human perturbations have already reduced—and in some areas eliminated—opportunities to study pristine habitats or communities within habitats.

Box 6: Interaction of anthropogenic effects and natural factors have led to complex changes in ecosystem structure and function.

SYNERGISTIC EFFECTS OF HUMAN ALTERATIONS AND NATURAL PROCESSES ON BIODIVERSITY

The Invasion of the Chinese Clam *Potamocorbula amurensis* in California:

The Chinese clam *Potamocorbula*, transported in ships' ballast water, was discovered in San Francisco Bay in 1986. Within several years densities reached 20,000/m^2, with the clam making up greater than 95 percent of the biomass, displacing much of the former benthic community. The invasion of *Potamocorbula* appears to have been facilitated by a 1986 flood that depressed the brackish northern Bay fauna, followed by the intrusion of high-salinity waters during an ensuing drought that prevented the return of the preflood fauna. *Potamocorbula* has persisted even after the return of normal salinities. *Potamocorbula* is in the process of completely altering Bay food webs: the clam has eliminated large summer phytoplankton blooms in the northern Bay. Clams continue to be supported in part by bacterioplankton. Cascading effects are expected as these trophic alterations affect zooplankton that, in turn, may impact fish populations.

Key References: Carlton et al. (1990); Nichols et al. (1990); Alpine and Cloern (1992); Werner and Hollibaugh (1993).

The Deterioration of Coral Reefs in the Florida Keys and Caribbean Basin:

Since 1982 natural factors enhanced by human activities have led to dramatic coral loss on many Florida and Caribbean reefs and to its replacement by algae. These natural phenomena include hurricanes, increased relative abundance of coral predators, and a massive pathogen-caused mortality of the algae-eating sea urchin *Diadema antillarum*. These events were greatly intensified because the reefs already had been altered by over-removal of herbivorous fish, magnifying the consequences of sea urchin loss. Eutrophication and nearshore development may have also tipped the balance in favor of algae. Additionally, coral loss due to extensive "coral bleaching" of uncertain origin—corals losing their essential, symbiotic algae—may be accentuated by eutrophication interacting with global warming and increased ultraviolet-B exposure. Reefs have proven resilient to local damage when the stress is ameliorated, but the cumulative and synergistic nature of all stresses may be irreversible on the scale of the duration of a human life.

Key References: Lessios (1988); Knowlton et al. (1990); Knowlton (1992); Porter and Meier (1992); Brown and Ogden (1993); Gleason and Wellington (1993); Jackson (1994); Sebens (1994).

Box 6: Continued

The Decline of the Seaweed *Fucus* in the Baltic Sea:

The status of the brown seaweed *Fucus vesiculosus* reflects the dramatic alterations that are now occurring in the abundance and distribution of life in the Baltic Sea. The *Fucus* community provides shelter, spawning, and foraging for many economically important fish. *Fucus* is now, however, greatly diminished: in southwest Finland it nearly disappeared in the late 1970s; near Kiel, Germany, *Fucus* is now found no deeper than 2 m with a 95 percent decline of biomass since the 1950s; and in Sweden it is now limited to 3-4 m depth having once occurred to 6 m in the 1940s. Eutrophication from human activities appears to be the basis for these changes, although synergistic interactions with natural upwelling may have led to the disappearance of *Fucus* in Finland. Eutrophication led to increased phytoplankton and thus decreased light penetration, and also stimulated dense growths of epiphytic algae on the *Fucus*, greatly reducing fucoid growth rates and increasing their drag, making the plants more susceptible to storm removal. In different regions the kelp *Laminaria saccharina*, the mussel *Mytilus edulis*, and several species of filamentous algae are now replacing *Fucus*. The Baltic Sea may be an ideal system in which to assess the consequences of distinct species replacements on ecosystem function because of its well-known hydrographic history and relatively low species diversity.

Key References: Kangas et al. (1982); Larsson et al. (1985); Kautsky et al. (1986); Launiainen et al. (1989); Vogt and Schramm (1991).

The Functional Extinction of Oyster Reefs in Chesapeake Bay:

A combination of the anthropogenic effects that are now widely altering marine biodiversity is demonstrated by the virtual elimination of the once vast oyster reefs of the Chesapeake Bay. The native oyster *Crassostrea virginica* once supported an enormous fishery that began to show significant declines a century ago due to overfishing and associated activities (such as the nonreplacement of shell for larval oysters to settle upon). Pollution closed local beds. Dredging, siltation, and marinas physically destroyed oyster habitat. Beginning in the 1950s, two disease agents, "MSX" (caused by the protozoan *Haplosporidium nelsoni*) and "Dermo" (caused by the protozoan *Perkinsus marinus*) led to the final demise of many remaining populations. These epidemics may represent resurgences of native species, or these protozoans may have been introduced. A similar fate with the same causes had earlier befallen the oysters of adjacent Delaware Bay. It has been calculated that Atlantic oysters in the Chesapeake Bay were historically abundant enough to filter much of the Bay's water once a week. Today the remaining oysters would require an entire year to do so.

Key References: Maurer et al. (1971); Hargis and Haven (1988); Newell (1988).

FISHERIES OPERATIONS

"'O Oysters', said the Carpenter,
'You've had a pleasant run!
Shall we be trotting home again?'
But answer came there none—
And this was scarcely odd, because
They'd eaten every one."

Lewis Carroll (1872)

Fisheries activities are especially ubiquitous across marine habitats, directly affecting virtually every habitat except the deepest sea floors. Even with management practices now in place, fisheries have major impacts on ocean environments, ranging from direct harvest to by-catch effects, habitat destruction, genetic changes, and food web changes. The challenge is to pursue a balanced view that incorporates ecological considerations and societal needs—that is, for long-term sustainable use of marine organisms for food and other products.

Direct Fisheries Effects

The extraction of large numbers of wild organisms (e.g., of finfish, sea urchins, seaweeds, shrimp, and scallops) from marine habitats through commercial fishing may be the most important impact of any current fishing activity. Cod and haddock (NOAA, 1992), Atlantic bluefin tuna (Safina, 1993), and many other fish populations of economic importance to the U.S. (Norse, 1993) have experienced dramatic declines due to overfishing. Both commercial and sport fisheries have intensively removed large populations of other edible, bait, aquarium, and curio trade organisms (such as mussels, abalones, limpets, clams, tropical seashells and corals, and polychaete worms) from coastal habitats. Little noticed when first surveyed by the present generation of scientists, most coral reefs were poised on the edge of profound change from overfishing of large predaceous and herbivorous fish (e.g., see Fig. 5).

Empirical and theoretical studies of trophic structure indicate that these removals do and will have a major effect on the composition and abundance of other species in these habitats and to the overall function of the ecosystem. Thus, on Georges Bank (United States/Canada), the original cod, flounder, and haddock fish populations, largely removed by fisheries, have been replaced by shark and ray populations, resulting in substantially different patterns of energy flow through the system (Backus and Bourne, 1987). In the Chesapeake Bay, a combination of overfishing, pollution, and eutrophication has greatly reduced the abundance of oysters and thus the role of the oysters in filtering the Bay's water (Box 6). Hunting of whales and seals in the nineteenth and twentieth centuries resulted in well-documented crashes of populations of these mammals (Evans, 1987), thus inevitably altering their roles in the food web.

Fisheries activities also cause changes in the demography of target species. Selectively removing the largest, and therefore oldest, components of the population can have dramatic effects. In the California sardine, for example, collapse ensued after the fishery had reduced the population from five to two reproducing year classes, and two consecutive years of poor environmental conditions led to spawning failure (Murphy, 1967). Extracting species with great longevity has often led to significant truncation in the age structure (Leaman, 1991), with potential genetic implications (Ryman, 1991; P.J. Smith et al., 1991). In tropical reef systems, a general pattern in the development of fisheries has been the selective removal of first the large species and then the large individuals within smaller species (Gobert, 1994). For species such as groupers with protogynous hermaphroditism, such size-selective removals have greater impacts on males, potentially altering the sex ratio, lowering mate supply, and thereby affecting the reproductive potential of the population (Shapiro, 1987). Fishing may change other life-history characteristics as well, resulting, for example, in earlier maturation and maturation at smaller sizes through the artificial selection regimes imposed on the populations (DeMartini et al., 1993).

Indirect Fisheries Effects

Three of the major indirect impacts of fisheries are by-catch—that is, capture and mortality of nontarget species (NOAA, 1992)—habitat destruction (discussed below, under "Alterations in Physical Habitat"), and ancillary impacts on interacting species or ecosystem effects. In the eastern tropical Pacific, tuna purse seine fisheries began in the late 1950s and incidentally encircled dolphins to capture the yellowfin tuna schools typically found underneath them. High mortality rates ensued for the dolphins, and populations declined by the mid-1970s to near 20 percent of the 1959 population estimate for eastern spinner dolphin and below 50 percent for the offshore spotted dolphin (T.D. Smith, 1983). Public concern about this by-catch was in part responsible for the passage of the Marine Mammal Protection Act. "Ghost fishing" (lost and abandoned nets that continue to capture fish and mammals) may have localized impacts. By-catch in marine ground fisheries can also have major impacts on fish (NOAA, 1992) and macrobenthos (Watling and Langton, 1994). Shell damage to clams has even been examined as an index of fishing activity in the North Sea (Witbaard and Klein, 1994).

Effects on abundance and demographics of interacting species likewise may be important; depletion of the great whales led to an increase in available krill by approximately 147 million tons per year (Laws, 1977, 1985). Marked changes were noted in growth rates and ages at sexual maturity in other marine mammals such as crabeater seals (Bengston and Laws, 1985) and southern minke whales (Kato, 1987). (For additional discussion of this topic, see Box 2.)

Mariculture Effects

Mariculture, the farming of animals and plants in the sea, has a long and rich history in coastal regions around the world, providing food and employment for many people. Fish, molluscan and crustacean shellfish, and seaweeds are among the most common organisms now cultivated. Large-scale, well-managed, and productive mariculture systems may further have the potential to preserve segments of marine biodiversity—at genetic, species, and habitat levels—by shifting attention away from the extraction and hunting of wild stocks.

Mariculture has, however, also had important impacts on coastal and estuarine habitats in many parts of the world (Ritz et al., 1989; Van der Veer, 1989; Tsutsumi et al., 1991; Rönnberg et al., 1992; Everett et al., in press). These impacts include extensive changes to the benthos and altered local nutrient inputs. Mariculture in the form of sea-ranching can reduce the genetic diversity of a species if the individuals in cultivation begin to contribute substantially to the gene pool of the wild population (that is, through external fertilization between gametes from cultivated and wild members of the species [Mork, 1991; Gharrett and Smoker, 1993]). Mariculture also has the potential to subject wild populations to an increased incidence of disease because of the well-known susceptibility of large, homogeneous populations (Lannan et al., 1989; Padhi and Mandal, 1994).

Effects of Scientific Collecting

The extraction of organisms from the sea by scientists, although on a much smaller scale than the other activities discussed here, may have local consequences. Scientists or their representatives may intensively collect selected species, a variety of species within certain habitats, or all species at a site, for environmental, ecological, or other biological studies. Scientific fervor over exceptional discoveries may have inadvertent consequences in this regard: intensive scientific collecting off South Africa and in the Indian Ocean of the coelacanth *Latimeria chalumnae*, a primitive bony fish, led to an increased rarity of this already rare fish (Bruton and Stobbs, 1991).

CHEMICAL POLLUTION AND EUTROPHICATION

Chemical pollution and eutrophication have altered the biodiversity of estuaries and coastal environments (Hughes and Goodall, 1992; Suchanek, 1993, 1994), and pollution has the potential to alter the biodiversity of deep-sea habitats through current and projected uses of this habitat for waste disposal.

The types and sources of marine pollutants vary widely (Hughes and Goodall, 1992). Organic and inorganic wastes enter the sea through sewage and industrial outfalls, river inputs (and their agricultural wastes), direct dumping, mariculture activities, spills and accidental losses of cargoes, and through atmospheric deposition of particulate materials. Halogenated hydrocarbons (pesticides, herbicides, and plastic compounds such as polychlorinated biphenols [PCBs]), heavy metals, petroleum products (including compounds such as polycyclic aromatic hydrocarbons [PAHs]), fertilizers (nitrogenous and phosphorous compounds), mining wastes, fuel ash, and radioactive materials are among the primary marine pollutants (Hughes and Goodall, 1992). An increased incidence of tumors and diseases in fish is one of the many consequences of contamination of estuarine and coastal environments by this broad range of pollutants (Myers et al., 1991; McCain et al., 1992; Vethaak and Rheinaldt, 1992).

Coastal eutrophication—nutrient enrichment from agricultural, sewage, and urban sources—has had severe impacts in shallow shelf areas and enclosed estuaries and bays worldwide (Nixon et al., 1986; Mannion, 1992; Turner and Rabalais, 1994). Macroalgal and phytoplankton blooms are frequent results which, in turn, often create conditions of hypoxia (low oxygen concentrations) and anoxia (no oxygen). Extensive invertebrate and fish mortalities may ensue (Norse, 1993). In particular, eutrophication has been linked to a more common occurrence of blooms of toxic algae (Hallegraeff, 1993; Smayda and Shimizu, 1993; Anderson, 1994).

ALTERATIONS IN PHYSICAL HABITAT

Coastal zones around the world have undergone significant physical alterations. In many regions large portions of salt marshes have been removed by dredging, filling, and diking to create dry land (Chabreck, 1988). Such activities are manifested today in tropical estuaries through the removal of mangrove communities for shrimp pond aquaculture (Robertson and Alongi, 1992; Norse, 1993). Long stretches of coastline in many regions of the world have been impacted by the emplacement of seawalls, jetties, groins, railroads, and other artificial structures that have altered natural patterns of sedimentation, erosion, and water flow. Mining has directly impacted intertidal and near-shore habitats and is a potential source of stress to the biodiversity of the deep sea. Upland and coastal mining, agriculture, and deforestation have caused extensive land erosion and the subsequent deposition of sediment, at times meters thick, in intertidal and shallow-water systems. In fact, sedimentation has become the major threat to certain coral reefs.

Other coastal habitats have been extensively altered through dredging (e.g., for ship channels) and by commercial dragging of the bottom in nearshore habitats for fish, clams, sea urchins, and other commercial targets (Matishov and Pavlova, 1994). Indeed, trawling and dredging on the seafloor are important indirect effects of fisheries operations. Recent surveys have revealed particularly profound impacts in the Gulf of Maine (Witman and Sebens, 1992; Watling and Langton, 1994) and elsewhere such as the North Sea (de Groot, 1984; see general review by Jones, 1992). In these regions the "megabenthos" (the largest bottom-dwelling animals and plants) have been entirely lost or significantly altered because of trawling and dredging activities.

By reducing the flow of water into estuaries, dams built throughout the twentieth century have significantly depressed the successful return of anadromous fish such as salmon and shad to their spawning grounds and have had many other effects on the local estuarine habitat by altering the natural salinity gradient (Skreslet, 1986). Although fewer dams are now being built, water diversion projects for agriculture and urban development are succeeding dams as a threat to estuarine biodiversity (Skreslet, 1986).

INVASIONS OF EXOTIC SPECIES

Biological invasions have become ubiquitous in virtually all habitats occupied or modified by human activities (OTA, 1993). Many estuarine and nearshore environments have been extensively invaded by exotic (nonindigenous) species, especially through the transport of larvae and spores in ballast water of ships, but also through introductions associated with mariculture. Box 7 focuses on biological invasions via ballast-water transport, and a specific example of ecosystem-level effects of a ballast-water invasion was given in Box 6.

The extent to which exotic invasions have affected the pelagic ocean, the coastal shelf, and tropical ecosystems such as coral reefs is largely unknown because of a lack of baseline information on the composition of these communities and a lack of studies focused specifically on invasions.

GLOBAL CLIMATE CHANGE

Atmospheric pollution is altering the exposure of the oceans to ultraviolet (UV) radiation and is increasing the concentration of gases that may lead to long-term climatic changes.

Compounds generated by human activities, including chlorofluorocarbons and brominated compounds rising into the stratosphere, destroy the ozone that shields the atmosphere from the sun's UV radiation. Increased UV-B (the biologically damaging UV) radiation penetrates many meters below the surface of the ocean (Fleishmann, 1989; R.C. Smith et al., 1992). UV-B exposure has increased under ozone "holes" in the Antarctic and elsewhere. In addition, recent satellite data indicate that volcanic activity (e.g., the 1991 eruption of Mt. Pinatubo) has reduced total air column ozone by as much as 10 percent (leading to increases in UV-B exposure of approximately 20 percent) in latitudes as low as Florida and the Bahamas (Gleason et al., 1993).

In turn, studies have confirmed that significant biological and ecological damage to phytoplankton and zooplankton (Hardy and Gucinski, 1989; Kramer, 1990; Behrenfeld et al., 1993a, 1993b), ichthyoplankton (Hunter et al., 1981),

---

Box 7: Exotic species introduced via ballast-water transport have displaced natural fauna and flora and may be threatening the world's coastal ocean biota.

BIOLOGICAL INVASIONS VIA BALLAST-WATER TRANSPORT

The most important global dispersal mechanism for passively moving shallow-water organisms between and across oceans is ships' ballast water and sediments. One estimate suggests that more than 3,000 species of coastal marine animals and plants are in transit around the world *at any given moment* in the ballast of ships. The result is that aquatic habitats all over the world are becoming dominated by exotic species. Scores if not hundreds of invasions have occurred during and since the 1980s alone. Examples include:

- The invasion of the Black and Azov Seas by the carnivorous American comb jellyfish *Mnemiopsis leidyi*, resulting in plankton biomass declines of as much as 90 percent and a startling decline in the anchovy fishery.
- The invasion of San Francisco Bay by the Chinese clam *Potamocorbula amurensis* (see Box 6).
- The invasion of the Texas coastline by the Indo-Pacific mussel *Perna perna*, which now forms monocultural stands on jetties for scores of kilometers along the Gulf of Mexico shore.
- The invasion of the Australian coast by Japanese red-tide dinoflagellates and by the carnivorous, abalone-eating Japanese starfish *Asteria amurensis*.
- The invasion of the Great Lakes by fouling zebra mussels (leading to economic losses of hundreds of millions of dollars per year), by three species of fish, and by a carnivorous water flea, all from European waters.

Introductions occur in all habitats, although they appear to be most common in estuaries and bays. Many invasions have profoundly altered the distribution and abundance of native species and the food webs of the systems they have invaded. Human introductions transcend natural dispersal barriers, bringing into contact organisms with no evolutionary experience between them, thus setting the stage for often dire results. The prospects for future spectacular invasions of coastal waters around the world remain extraordinarily high, as long as ballast water continues to be moved and released.

Key References: Carlton (1985, 1989); Shushkina et al. (1990); Carlton and Geller (1993); Nalepa and Schloesser (1993).

---

and benthic organisms (Jokiel, 1980; Jokiel and York, 1984; Kramer, 1990; Bothwell et al., 1994) from UV-B can and does occur in relatively shallow or surface waters. Recent work has further documented the effects of increased UV-B on corals (Gleason and Wellington, 1994). The species-specificity of Antarctic phytoplankton susceptibility to UV-B damage, for example, suggests potential changes in the size and taxonomic structure of the phytoplankton assemblage (Karentz et al., 1991).

The burning of fossil fuels and global agricultural practices have increased the amount of carbon dioxide, methane, and other gases in the atmosphere—such gases trap the heat radiating from the Earth, creating a "greenhouse effect." Continued increases in these gases have a strong potential to lead to global warming, and many scientists think that such warming has begun. A warming Earth could affect the sea, from the most inland marshes to the deepest oceans, in predicted ways that range from sea-level rise to modified patterns of rainfall and oceanic circulation (which, in turn, would affect nutrient supply and distribution). Increased sea water temperatures may alter the abundance, distribution, and reproduction of many coastal species (Ray et al., 1992), and may make northern regions more susceptible to invasions by warm temperate and subtropical species (Chapman, 1988).

As with all other environmental perturbations, there exists the potential for synergisms and cascading effects of global climate change that have not yet been considered, and which may interact with the other stresses reviewed here.

FOUR

Regionally Defined Model Systems: Examples of Habitats

As has been stated previously, the processes responsible for biodiversity changes due to anthropogenic effects must be studied on all *relevant* scales. One of the most important results of metapopulation theory for the conservation of biodiversity is that changes in the abundance of a species at some sites may cause its extinction or increase at other sites as an inevitable result of its vital statistics, natural history, and colonization potential (Nee and May, 1992). Thus conservation and management of marine biodiversity ultimately depend on understanding regional processes. Such a regional approach must simultaneously include relevant small-scale ecological studies and good taxonomy, which are essential to the entire picture. These activities must fit together to provide the necessary links between local and regional perspectives.

Much of the difficulty in studying regional biodiversity is that it depends on both ecosystem and population processes that are traditionally studied at different scales and in different disciplines (S.A. Levin, 1992). On the one hand, biological oceanographers have tended to emphasize ecosystem processes such as primary production operating on the scale of ocean currents and upwelling systems. On the other hand, marine ecologists have tended to emphasize processes affecting populations of species on a local scale (e.g., patches, see Figures 1 and 2). *Thus a major goal of this marine biodiversity initiative would be to bring together population and ecosystem approaches at the same scales to understand, when possible, the origin and maintenance of marine biodiversity.*

These considerations, along with the rationales discussed in Chapter 2 "Linking Pattern to Process," suggest that it will be necessary to focus on regionally defined model systems. These systems are of the appropriate scale to address questions concerning the linkages between ecological processes operating at relatively small scales and oceanographic processes impacting these systems at larger scales. A dedicated regional focus provides the framework for simultaneously conducting basic and applied research in biological oceanography, ecology, and conservation biology aimed at understanding changes in biodiversity due to human activities.

Discussed below are six types of habitats that would be contained within regional model systems. They are described in terms of the intrinsic interest and value of the regions, how they have been or may be impacted by anthropogenic effects, and the compelling opportunities for study. The systems cover the complete range of spatial scales discussed earlier (see also Figure 3), but the degree of understanding of processes that control biodiversity within and across these scales varies widely among the systems. The first three systems (estuaries and bays,

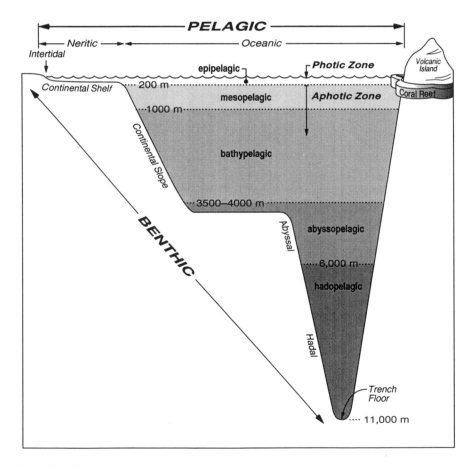

FIGURE 3 Regions of the marine environment.

coral reefs, and temperate rocky shores) have been and are undergoing heavy impacts from human activities. For the last three systems (shelf-slope systems, the deep sea, and the pelagic open ocean), a sense of urgency relative to what human activities may now be doing or could do in the near future is driven by profound ignorance of how diversity in these systems is created or maintained.

ESTUARIES AND BAYS

The majority of people in the world live within 100 kilometers of bays and estuaries (Norse, 1993), and such environments command enormous attention and use. Estuaries have long been associated with:

• *some of the world's greatest fisheries for oysters, clams, shrimp, crabs, and fishes.* In addition, bays, marshes, and seagrass beds act as well-known "nursery grounds" for larvae and juveniles of many species (Weinstein, 1979; Boehlert and Mundy, 1988). Yet many of these notable fisheries are at or near collapse because of overfishing, disease, and pollution (e.g., Maurer et al., 1971; Newell, 1988).

• *marshes, seagrass beds, and mangroves.* Such coastal vegetation is ecologically critical as detrital and nutrient sources driving nearshore production, filters for land runoff, protection from coastal storms, sediment traps, and sediment stabilizers (Fenchel, 1977; Adam, 1990). Marshes also have important aesthetic and recreational value in their "pristine" condition (Teal and Teal, 1969; Chabreck, 1988). They are being rapidly lost, however, because of extensive draining and filling practices (Dahl et al., 1991).

• *harbors and marinas.* Dredged channels maintain open routes for economically important international shipping. In turn, international shipping means that ballast water is discharged frequently, facilitating the establishment of exotic species (see Boxes 6 and 7) that can significantly change ecosystem structure and function (e.g., Fig. 4). Moreover, floats, docks, and pilings have replaced mudflats, marshes, and seagrass meadows to provide space for recreational pleasure boats (Zedler, 1994).

• *pollution discharge.* Estuaries and bays have long received the bulk of human-generated municipal and industrial wastes that enter the oceans from the land, rendering many of them unfit for fisheries production (Kennish, 1992; Schubel, 1994).

These and many other activities in the world's most populated areas adjacent to marine environments mean that estuaries and bays are where the greatest proportion of natural habitat has been destroyed or severely altered (Dahl et al., 1991). Thus, estuaries and bays, and their component marshes, seagrass beds, and mangroves provide striking opportunities for understanding how a profusion of human activities, acting singly and in combination, decrease, maintain, or increase biodiversity (e.g., see Box 5). In many estuaries, the original life is all

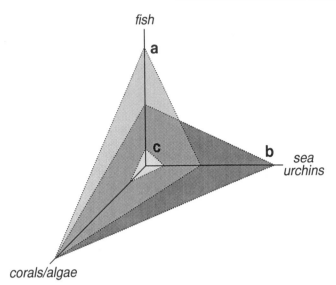

FIGURE 4 Changes in chlorophyll *a* concentration with increasing abundances of the introduced clams *Potamocorbula amurensis* and *Mya arenaria* in San Francisco Bay. *Source*: Alpine and Cloern (1992).

but gone, in terms of species diversity, abundances, and local distributions. In more remote estuaries, little is known regarding the balance between pristine and altered conditions. And, for other estuaries, there appear to be far more species now than ever before (e.g., 200 introduced marine, brackish, and freshwater animal and plant species have been added in 150 years to the San Francisco Bay and delta region [Carlton and Cohen, pers. comm., 1994]), but little is known about the structural or functional consequences of this increased diversity, or of the addition of so many species to a single system at such a rapid rate.

In addition, the naturally lower diversity of most estuaries provides an opportunity to study the relationship between diversity and ecosystem function over a range of systems (see Box 9).

CORAL REEFS

"At the rate things are going, there may be very few Caribbean corals to be affected by global climate change (Smith & Buddemeier 1992)."

Jackson (1994, p. 59)

Coral reefs are renowned for their remarkable diversity (Connell, 1978). Reef organisms display extraordinary specializations, intensive predator-prey evolutionary pathways, and competitive interactions within and among species (e.g., Jackson, 1991). Because this diversity is expressed in a spectacular array of form and color, and because of increasing threats to the survival of this diversity (Hallock et al., 1993), coral reefs have attracted considerable tourism and public interest. Moreover, modern coral reefs, representing approximately 6,000 years of growth during the most recent period of sea-level rise, are the oldest and largest biogenic structures in nature bearing detailed paleoecological and climate records (Dunbar and Cole, 1993; S.V. Smith and Buddemeier, 1992). Whereas global climate change, and the potential associated effects of sea-level rise, increasing water temperature, and increasing ultraviolet (UV-B) radiation, are possible long-term threats to reefs, there is an immediate need to manage the relentless impact of explosive human population growth on reef habitats (D'Elia et al., 1991; Norse, 1993).

Reefs provide good examples of the importance of linkages between habitats—reef biodiversity is dependent on adjacent ecosystems for feeding areas and nursery grounds and as buffers against land runoff of sediments and nutrients (Kuhlmann, 1988; Ogden, 1988). Runoff from terrestrial environments is one of the most severe problems affecting reefs around the world. Sediment load from agriculture or forest clear-cutting, freshwater, and industrial activity are known to be damaging reefs (Grigg, 1984, 1994; Hodgson, 1989; Richmond, 1993; Sebens, 1994) either by killing colonies, preventing settlement of new recruits, interfering with sexual reproduction, or all three. Eutrophication from agricultural fertilizers and human sewage is a particular problem, *first* because organic enrichment causes faster growth in weedy macroalgae than in corals, overgrowing and killing them (S.V. Smith et al., 1981), and *second* because corals are adapted to live in a nutrient-poor environment, and thus overfertilizing alters the productive but very fragile relationship between the corals and their algal symbionts (Falkowski et al., 1993). In addition, reefs hold a significant portion of the fisheries resources of developing tropical countries, and they are very sensitive to overfishing, especially the removal of large predators and herbivorous fishes (see Boxes 2 and 6).

Biotic and abiotic disturbances on reefs that shift reef composition from framework builders (corals) to nonframework builders (algae) have particularly dramatic effects on biodiversity. One example is provided in Figure 5 that illustrates how the delicate natural balance (a) between fishes, the ratio of coral to large algae, and the herbivorous sea urchin *Diadema antillarum* in Caribbean reefs has been tipped first one way (b) by overfishing and then another way (c) by mass mortality of the urchin due to an unknown pathogen (Jackson, 1994).

Excellent opportunities exist in coral reef systems to look directly at the dynamic interface between the natural patterns, processes, and consequences of biodiversity, and the increasing pressures from human activities. A broad range of observed transitions between different reef communities exhibiting differing levels of impacts and thus threshold effects—effects that may be irreversible over the scale of a human lifetime (Knowlton, 1992)—offers an irresistible comparative menu for study.

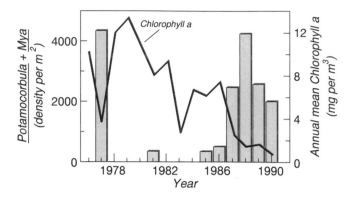

FIGURE 5 Hypothesized response of Caribbean coral reef communities to overfishing and sea urchin disease. *Source:* Jackson (1994).

TEMPERATE ZONE ROCKY SHORES

Rocky intertidal zones, and the shallow sublittoral kelp forests just below them, have provided the conceptual framework for research on most other benthic ecosystems (Connell, 1961; Paine, 1980). They therefore offer one of the best possible opportunities for rapid advance in understanding of the dynamics of regional systems.

Many temperate rocky shores have been very extensively altered by the virtual removal of large vertebrate predators such as sea otters (Estes and Palmisano, 1974). Human foraging on intertidal shores and, in more recent decades, recreational tidepool tourism have resulted in some of the more dramatic demonstrations of how direct removal of large numbers of invertebrate and vertebrate carnivores and herbivores, combined with sustained trampling, can locally obliterate the fauna and flora of a rocky shore (Beauchamp and Gowing, 1982; Bally and Griffiths, 1989). Thirty years of visits by hundreds of thousands of school children to the tidepools of what is now a "marine reserve" in central California have resulted in the complete local absence of seashore snails and crabs that are considered to be the most typical representatives of Pacific marine life (R. Breen, pers. comm., 1994). In Chile, the best experimental examples of the devastating effects of human extraction of animals on a marine community come from the rocky intertidal zone (see Box 13).

Rocky shore systems, because of their geographic ubiquity and experimental tractability, remain prototype systems for exploring the connection between biodiversity and ecosystem function. Biodiversity issues at all trophic levels, especially at the microbial level, remain unexplored. The interaction between nearshore oceanography and larval transport is basic to understanding the presence, absence, and distribution of species. This interaction, set against the high variability of dispersal capability among species, seems certain to generate variation in community composition at large spatial scales. The extent of this variability, its relation to whether particular sites are larval sources or sinks, and biodiversity as a consequence of local interaction remain major topics in metapopulation dynamics and conservation biology. Marine, nearshore conditions provide an appropriate environment to investigate the interplay between connected populations in open systems.

CONTINENTAL SHELVES AND SLOPES

"The canneries themselves fought the war by getting the limit taken off fish and catching them all . . . It was the same noble impulse that stripped the forests of the West and right now is pumping water out of California's earth faster than it can rain back in."

Steinbeck (1954)

Continental shelves represent the great interface between the continents and open oceans. Directly and indirectly impacted by the natural and human effects that operate on land margins, the shelves are the only "open ocean" most people will ever see or know—and yet the shelves are separated by boundary currents from most of the sea. The concept of shelf waters representing "the ocean" is reinforced by the popularization since the 1950s of *"food from the sea,"* resulting in shelves being portrayed primarily as sites of most of the major world fisheries, located on fishing banks, in upwelling zones, and on broad shallow platforms.

Shelf waters have suffered habitat alteration and changes in biodiversity due to overfishing and the extensive physical damage caused by the deployment of mobile fishing gear (Graham, 1955; Hutchings, 1990). Regional "hot spots" include dramatic shifts in community structure in northeast American shelf fish communities (NOAA, 1992) and the Eastern Bering Sea shelf, where a critical environmental concern is the decline in abundance of certain marine mammal and seabird populations (Alverson, 1992; Pascual and Adkison, 1994), with the Steller sea lion being particularly threatened. Increasing stresses in offshore systems from inshore pollution are of escalating concern but remain poorly understood. Other stresses arise from oil spilled in the sea during tanker transport and operations (NRC, 1985), and to a lesser extent, from oil and gas exploration on the "outer continental shelf," the subject of extensive environmental impact studies since the 1960s (NRC, 1985; Boesch and Rabalais, 1987).

The seemingly distant and more immune slope waters are no longer far away and no longer immune. Deep-water fisheries—and their attendant physical effects (e.g., habitat alteration due to dredging and trawling)—have entered slope waters (Robertson and Grimes, 1983; Polovina and Ralston, 1986). Deep oil and gas mining will soon commence on slopes at and below 2,000 meters in the Gulf of Mexico (R. Carney, pers. comm., 1994).

Shelf-slope systems thus provide abundant opportunity for examining the impact of anthropogenic activities—of which fisheries-related effects have been and are the most conspicuous—within a regional-scale context. The extent of spatial heterogeneity of shelf-slope systems requires particular attention, striking as it does at the heart of assumptions in ocean management about seemingly "large and monotonous" habitat systems. In particular, research approaches should be undertaken that examine the contribution of several different areas of the shelf to the recruitment of key species, specifically incorporating new techniques for larval tracking (L.A. Levin, 1990; L.A. Levin et al., 1993; see also Boxes 11 and 12). Shelf waters also provide opportunities to study multiple steady states in community composition, the reversibility of diversity shifts, and how long such reversals could or do take, species redundancy, and altered energy budgets.

PELAGIC OPEN OCEAN

The open oceans of the world cover nearly three-quarters of the Earth's surface and have been studied for well over 100 years; yet, there are continuing discoveries of new higher taxa and new communities. These include a recently discovered group of microorganisms that in some areas may contribute almost as much to oceanic primary production as all previously known primary producers (Fig. 6; Chisholm et al., 1988, 1992; R.J. Olson et al., 1990; S.W. Chisholm, pers. comm., 1994). The pelagic ecosystems that contain the highest diversity are the epipelagic and mesopelagic strata between the surface and approximately 1,000 meters, within which vertically migrating plankton and micronekton provide active biological linkages.

The public's view of the open sea as a vast homogeneous body of water is belied by the complexity of oceanic subsystems: fronts and eddies, upwelling and downwelling regions, boundary currents, and large ocean gyres (Longhurst, 1981). Such subsystems offer exciting opportunities for the study of biodiversity patterns and the processes that maintain them (Angel, 1993). In regions of intensive fishing pressure over large spatial and temporal scales, for example, there are unique opportunities to investigate "top-down" controls where sufficient historical data are available (discussed in Chapter 6, "Retrospective Analysis: Importance of an Historical Perspective").

Unlike other marine and terrestrial ecosystems, the pelagic realm is not dominated by substrate-related processes. Thus, the ecosystems of the open oceans provide a test of the generality of paradigms developed elsewhere, the resolutions of which are essential for a balanced understanding of marine biodiversity. For instance, studies from some substrate-dominated ecosystems have led to the paradigm that biodiversity is directly related to environmental hetero-

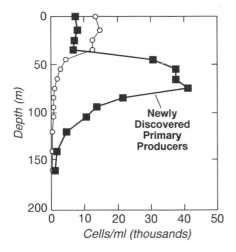

FIGURE 6 Total numbers of prochlorophytes (squares), a newly discovered group of marine, free-living, primary producers in the picoplankton, greatly exceeded those of the other group of primary producers in this size range, the cyanobacteria or blue-green algae (circles), in samples from the euphotic zone off of southern California. *Source:* Chisholm et al. (1988)

geneity (e.g., Tilman and Pacala, 1993). Yet, in the open ocean, the most diverse communities occur in environments of very low physical variability (McGowan and Walker, 1993). Likewise, studies of intertidal and coral reef ecosystems have shown the importance of predation in maintaining the diversity of the prey species ("top-down" control) (e.g., Paine, 1980, 1992; Jackson, 1994). This has not been demonstrated in open-ocean ecosystems, which are usually assumed to be controlled by nutrient input ("bottom-up" control). In fact, some of the most exciting recent, large-scale experimental research in the oceans has been on the role of iron as a limiting element controlling primary productivity (Ditullio et al., 1993; Martin et al., 1994).

The fluid nature of pelagic environments shifts natural variability towards slowly developing, long-lived fluctuations. Many biological interactions occur over short temporal scales which, when coupled with the large spatial extent of many pelagic environments, result in subsystems that are similar in all but a few scales of environmental processes. Comparisons between such systems can provide information akin to that derived from very large manipulative experiments. For instance, the eastern and western basins of the North Pacific central gyres may be contrasted to examine the influence of increased mesoscale variability and consequent nutrient input on the biodiversity of the western basin. Comparison between the subarctic gyres of the Atlantic and Pacific may show how the depth of winter mixing, which is known to influence the composition of the primary producers, is transmitted through the ecosystem—another scale of "bottom-up" control.

Many open-ocean ecosystems have, however, been less influenced by recent anthropogenic changes than have their coastal counterparts. Because of their relatively pristine nature and the relatively low levels of natural short-term fluctuations, these pelagic systems potentially could provide some of the first opportunities to evaluate ecosystem responses to the slowest, but most pervasive, effects of human activities, such as global warming, increased ultraviolet radiation, and the gradual accumulation of pollutants. In order to recognize, and ultimately to predict, such responses, it will be necessary to identify the functionally significant organisms in the pelagic ecosystems and to establish their responses to the anticipated anthropogenic agents. Therefore the open ocean should provide excellent opportunities to relate biodiversity concerns, traditionally the domain of population biologists and taxonomists, to issues of ecosystem organization and function.

DEEP SEA

"The frequency with which the deep ocean is popping up as a fanciful solution to waste and resource problems is extremely worrisome. I see four origins for this trend. First, it is the product of our success in defending better known habitats like reefs and bays. Second, it is the persistent belief among non-biologists that the abyss is the desert proposed by Issacs (seep and vent communities are still referred to as oases). Third, the distant deep ocean is the ultimate 'not in my backyard.' Finally, the vastness of the abyss gives rise to a wasteland management policy which views insults to relatively small areas as fully acceptable."

R. Carney (pers. comm., 1994)

The number of species contained within the deep sea—the least-studied marine habitat—has been a subject of recent, intensive debate (J.F. Grassle and Maciolek, 1992; May, 1992; Poore and Wilson, 1993). There is no question, however, that the previous notion of a global deep-sea bottom that is uniformly featureless has been shattered over the last two decades by countless discoveries of unique, sometimes bizarre, and highly diverse deep-sea communities (e.g., Butman and Carlton, in press). Each major ocean basin has a distinctive fauna, and bottom assemblages vary according to latitudinal gradients (Rex et al., 1993) and topographic features such as basins, canyons, and areas of strong currents (J.F. Grassle, 1989). Discrete, ephemeral patches of food (Billett et al., 1983; Wolff, 1979; Suchanek et al., 1985; Gooday and Turley, 1990) or biogenic structure (C.R. Smith et al., 1986; Thistle and Eckman, 1990), and defaunated patches produced by disturbance (C.R. Smith and Hessler, 1987; Snelgrove et al., 1994), are common and introduce a measure of small-scale spatial heterogeneity. At larger spatial scales, disturbances generated by upwelling regions, episodic strong currents (bottom boundary currents, canyons), slumping (trenches or steep slopes) or heavy sedimentation (e.g., Mississippi plume, or offshore of Cape Hatteras) introduce large-scale patterns (greater than 1,000 square kilometers) that appear to obliterate the patterns of discrete, widely separated, ephemeral patches (J.F.

Grassle, 1989; Schaff et al., 1992; Blake and J.F. Grassle, in press) or gradual community changes with depth (Carney et al., 1983) and sediment grain-size diversity (Etter and J.F. Grassle, 1992).

Although long considered inaccessible and difficult to sample, better quantitative sampling and improved methods of experimentation are continually improving the ability to study the deep-sea biota (see Box 12). The surprisingly high diversity of benthic invertebrates in the deep sea—hundreds of species co-occurring within a square meter of ocean floor—provides a remarkable platform from which to renew empirical, experimental, and theoretical attempts to explain the causes and patterns of global diversity (Rex et al., 1993).

Although the deep sea is vast and remote, humans are still having an effect on this environment. A small, steady increase in abyssal temperature of 0.32°C in 35 years has been attributed to global climate change (Parrilla et al., 1994). Increases in pollutants from atmospheric sources have been observed on the deep-sea floor (La Flamme and Hites, 1978; Takada et al., 1994). Ocean dumping (waste disposal) and deep-sea mining are clear causes for concern. Changes in upper-water column trophic structure—directly or indirectly, for example, through pollution, overfishing, or global climate change—potentially affect deep-sea trophic dynamics (e.g., see Box 1). Because time-series information on biological diversity is unavailable for any deep-sea area, unknown changes currently may be in progress, and thus, the deepest oceans of the world are not immune to human actions. Underscoring these actual and potential changes is the concern that anthropogenic effects on deep-sea communities may be especially severe because deep-sea organisms may be more poorly adapted to change (J.F. Grassle et al., 1990).

Increasing technological sophistication for studying the deep-sea environment (see Box 12) will permit biodiversity issues to be examined experimentally at traditional, necessarily small spatial scales (cm^2 to m^2). Deep-sea mining technologies and waste-disposal experiments can be designed to produce well-replicated and well-controlled experiments at scales of tens to hundreds of km^2. The integration of these approaches should promote better understanding of how and which processes generate and maintain biodiversity in a global environment containing one of the richest species complements.

FIVE

The Critical Role of Taxonomy

Changes in the sea caused by anthropogenic effects are most commonly measured by changes in the distribution and abundance of species. The loss of individual populations of species may affect the genetic diversity of a species, and thus impact the survival of the species itself. The loss of ecosystem diversity restricts the habitat available for a species, and so, too, may affect the species' survival. At the center of these cascading effects is the *species*. The ability to identify individual species is thus the key that permits the opening of the first door to an understanding of community structure and function. And yet for all marine systems, the ability to "simply" identify the species present is now threatened by a continuing loss of scientists with the knowledge and ability to understand and describe biodiversity. Moreover, in many systems, species diversity is so poorly known—that is, so many species and entire groups of higher taxa remain undescribed (see Table 1)—that the impact of human activities on diversity is difficult to assess at all.

Even documenting the most obvious patterns of change is proving to be surprisingly difficult. In many cases, there are no or an inadequate number of trained taxonomists in a geographic region to allow the documentation of the distribution and abundance of even well-known species. Furthermore, the description and understanding of new species, genera, families, or even phyla are hindered by declining numbers of taxonomists (Stuessy and Thomson, 1981; SA2000, 1994) and the poor dissemination of taxonomic information to other field and laboratory biologists.

Most taxonomists describe species and undertake major taxonomic revisions according to their own professional interests with little or no communication with

---

Box 8: Training in taxonomy has declined dramatically over the last several decades, so that the appropriate scientists may not be available to describe the next round of unique species in a newly discovered marine habitat.

WHO DESCRIBED THE HYDROTHERMAL VENT FAUNA?

All but a few of the taxonomists who described the hydrothermal vent fauna, beginning in 1979, were over 40 years old at the time they published. In the mid-1990s, the *same* taxonomists—those who have not retired—continue to describe organisms from hydrothermal vents. These individuals were specifically chosen to describe the vent fauna because of their exceptional taxonomic expertise. Those who are still working continue to contribute substantially to the taxonomy and systematics of the group of animals in which they specialize, and some are the *only* remaining experts on their particular group of organisms. In fact, for most taxonomic groups, there has been little or no training of younger workers in the identification and description of species.

When the next major novel ocean ecosystem is discovered, who will describe the animals, plants and microbes associated with it?

Key References: Jones (1985); Tunnicliffe (1991).

---

ecologists. At the same time, it is becoming increasingly apparent that a "graying" of taxonomists is occurring (e.g., Box 8). Taxonomists specializing in taxa of ecological importance have retired or died and there has been little or no training of younger scientists in the field; consequently, ecologists are being left on their own to deal with the seemingly bewildering array of species occupying most ecosystems. Unfortunately, most ecologists do not have the training to deal adequately with this task. Few ecologists today have had any formal training in taxonomic methods and principles, or in the detailed morphology of the group(s) that they study, and even fewer are aware of the ecological importance of such basic knowledge.

There are many examples of ecological studies that have been compromised by taxonomic mistakes (some examples are described in Lee et al., 1978; Knowlton, 1993; Knowlton and Jackson, 1994). Yet in contrast to a strong movement toward requiring, for example, rigorous statistical designs and analyses in ecological studies (with some number of submitted manuscripts being rejected by editors on the basis of inadequate statistical treatment), there are currently no rewards or penalties for good or bad taxonomic work on the part of ecologists and biological oceanographers, nor clear mechanisms by which to assess the quality of such work. Nevertheless, taxonomic competence is just as important for ecology as are rigorous statistics.

Fundamentally, however, the taxonomic bottleneck will only be passed when taxonomists and ecologists begin speaking the same language. This means that taxonomists will need to receive more training in the methods of ecologists, and ecologists will need to learn more taxonomy. Taxonomists, on their own or working with ecologists, can and should use ecological and genetic methods to sort out cryptic sibling species. A recent example of such work is the recognition of cryptic sibling species in mussels of the genus *Mytilus*, where molecular genetic studies were combined with classic taxonomic approaches to resolve the species complex (see "Significant Opportunities for Forging New Horizons" in Chapter 1).

Likewise, ecologists, working on their own or with taxonomists, can and should describe or revise descriptions of species discovered as a result of their ecological research. Examples of the latter are the joint description by a taxonomist and an ecologist of a new species of isopod crustacean living cryptically on a tropical bryozoa (Buss and Iverson, 1981), and a new species of galatheid crab from hydrothermal vents (Williams and Van Dover, 1983). New methods in taxonomy and systematics, including molecular and cladistic techniques, have made these tasks easier, and have revealed a host of important ecological differences between closely related species.

Do we have to know the name of every species in a given region to address human impacts on biodiversity? It is clearly impractical to locate, identify, and describe every living organism in every community. The level of investigation of biodiversity should be dictated by the basic scientific questions being asked, and the urgency of the perceived environmental threats that demand the development of more informed policies.

Biodiversity Research Program

CONCEPTUAL FRAMEWORK

In order to describe, understand, and predict changes in marine biodiversity, well-defined, integrated research questions must be addressed. These questions must be formulated in light of the critical environmental issues prevalent in nearly all marine habitats. Furthermore, the questions must be asked across the large spatial and temporal scales defined by the open nature of most marine systems (see Box 4). The appropriate geographic and temporal scales must be explicitly defined for each regional system under consideration. Scales are in part defined by the strength of the linkages between localities, and these linkages can be determined using newly developed genetic and survey techniques (see Boxes 11 and 12) as well as through studies of the coupling of biodiversity patterns over large spatial scales (such as geographic variation in the intensity of larval settlement [Caffey, 1985; Ebert and Russell, 1988]).

Such scales tend to transcend limits normally set in biological research efforts. But traditional limits must be surmounted if the effects of anthropogenic activities on biodiversity are to be meaningfully understood in marine systems. Research tools to address biodiversity issues over the appropriate marine scales are now becoming available because of significantly improved understanding of many processes that control marine biodiversity at local scales and because of the development of new techniques for species identifications, genetic description, habitat sampling, and conducting experiments in the oceans (see Boxes 11 and 12).

For any experimental system, the questions must focus on a hierarchy of different levels: (1) patterns of biodiversity; (2) anthropogenic and natural pro-

---

Box 9: Differences in environmental heterogeneity and biodiversity within and between regions provide a foundation for formulating biodiversity research questions on the patterns of diversity and why diversity matters.

POLAR, PELAGIC, AND TEMPERATE-SHELF ECOSYSTEMS: CONTRASTS IN DIVERSITY AND ENVIRONMENTAL HETEROGENEITY

Three regional systems—representing polar, pelagic, and temperate-shelf environments—offer tantalizing contrasts in their relative diversity and ecosystem structure. They serve as examples of the types of systems within and between which compelling research questions on the patterns of and the processes controlling diversity could be formulated:

• One of the most interesting contrasts in polar marine ecology is between the Antarctic and Arctic, with the former having a much higher species richness (for example, four times the number of mollusks) than the latter. Yet the Antarctic lacks the ecological (habitat and physical) diversity of the Arctic. Historical processes have led to some of the observed differences—the Arctic is heavily disturbed and has a younger fauna—but explanations for many differences in diversity and ecosystem function remain illusory.
• The Baltic Sea has far lower diversity than the adjacent North Sea, due primarily to differences in salinity. Despite this contrast, there appear to be few if any striking differences in energy production and flow in the water column and benthos. Both seas have similar major functional types of organisms (such as benthic macrophytes, phytoplankton, and suspension-feeding clams and mussels). What role does this level of similarity play in apparently reducing the role of diversity in energy dynamics?
• In the central Pacific Ocean pelagic ecosystem, the western and eastern portions appear to be comparable in species composition and structure. However, there is a significant increase in nutrient input in the eastern portion. This difference in frequency of nutrient injection may provide an excellent opportunity to determine how one scale of environmental heterogeneity does—or does not—influence biodiversity dynamics.

Key References: Elmgren (1984, 1989); McGowan and Walker (1993); Dayton et al. (1994).

---

cesses that generate or alter these patterns, and natural processes that historically generated a given pattern; and (3) consequences to ecosystem function of biodiversity change (e.g., Box 9).

Patterns

Adequate knowledge of patterns of biodiversity is basic to understanding and predicting the processes responsible for these patterns. Patterns include changes in biodiversity over time and space. They include not only the species found in an area but also their relative abundances, genetic diversity, and apportionment in higher taxa. Pattern data are critical in understanding the processes that affect biodiversity and in detecting biodiversity changes.

Processes

Processes that determine or alter biodiversity have been studied extensively in some habitats in the sea, but generally over scales too small to allow understanding of entire marine ecosystems. Furthermore, the blending together of human impacts and natural ecological processes in pattern formation has been slow to be accepted as a theme in traditional marine ecology. Yet, successful, intensive, local studies of marine systems (e.g., the rocky intertidal; Paine, 1980; Paine and S.A. Levin, 1981) serve as critical signposts to guide regional studies of biodiversity. Moreover, retrospective studies have the potential to identify natural processes that historically generated a given pattern. These signposts point towards questions centered on critical environmental issues. When do different marine species play interchangeable ecological roles? To what extent do human activities make species or communities more susceptible to natural pressures like disease, physiological stresses, or habitat destruction? When do increasing human impacts push ecosystems across ecological thresholds? Are precipitous declines of marine species a result of these thresholds?

Consequences

Why does biodiversity matter? In addition to a moral and aesthetic imperative, decreasing or otherwise changing biodiversity in marine systems may have important economic effects (e.g., the collapsing fisheries around the world [Norse, 1993]). What are the most important practical reasons for maintaining natural levels of biodiversity in marine systems? This kind of question can only be answered by understanding the functional significance of biodiversity in marine ecosystems—in terms, for example, of how species diversity influences production, of how genetic diversity influences population growth or epidemics, or of how natural diversity levels confer resistance or susceptibility to invasions or to the ability of a system to recover from natural and human impacts (e.g., Tilman and Downing, 1994).

These relationships are basic to understanding diversity in all environments, not just marine. In some cases, marine ecosystems provide the best experimental platforms for testing the ecological correlates of diversity in open systems. Thus, the scientific questions to be addressed would have important implications for the study and understanding of biodiversity throughout the world, providing a synergism between the unique aspects of marine and terrestrial initiatives.

BASIC RESEARCH QUESTIONS

Listed below are some important overall questions about marine biodiversity that would constitute a major focus of this initiative. For any given marine environment, it is impossible to answer every question. However, various marine systems represent opportunities to address these questions in different combinations and ways. Fundamental advances in understanding the biodiversity of marine systems will come from coordinated studies of individual ecosystems at all three levels outlined above: pattern, process, and consequence. To apply broadly to an open system, these studies must be conducted over the scale appropriate for that system, and the results must be integrated into a dynamic view of biodiversity patterns over space and time.

Because, in this initiative, the biodiversity questions would be addressed within the context of anthropogenic effects, the end result of these studies would provide critical information about practical problems in marine ecosystems, how these problems act at the ecosystem and local levels, and how connectivity between different localities may increase or buffer biodiversity change. *Research conducted within this initiative is thus designed to be of a breadth, depth, and scale such that a more complete environmental picture can emerge, facilitating many potential management and conservation applications.*

Natural Variation in Biodiversity Pattern and Why Biodiversity Matters:

1. How do genetic, species, and ecosystem diversity vary in space and time at different regional scales, and within habitats within these regions? Examples of specific research questions are:

• To what extent does the maintenance of local biodiversity (genetic or species) depend on linkages between distant populations, the dispersal between them, and the availability of suitable habitat?
• How does genetic diversity within a species influence reproduction and population growth or susceptibility to epidemic disease?
• To what extent do changes in biodiversity at one site within a region—or between regions—affect the biodiversity at another site or in another region?
• What specific characteristics of a habitat directly or indirectly influence genetic and species diversity? For example, are there parallels in the origin and maintenance of coral reef and deep-sea biodiversity?

2. What is the functional significance of biodiversity at the genetic, species, and ecosystem levels? Are species within a functional group interchangeable? What might be learned from comparing and contrasting systems in terms of the functional significance of biodiversity? (For example, are there parallels between the ecological significance of microbial diversity as coral reef symbionts [zooxanthellae] and as open-ocean primary producers [picoplankton]?)

3. To what extent does the diversity of a community determine (a) "stability," (b) productivity, (c) resistance to invasion or disease, and (d) ability to recover from natural and human impacts? Equally important, how do these factors interact? Do high diversity systems have higher or lower production than systems whose diversity has been impaired? What is the role of biological invasions in altering system production or energy flow?

4. How good are the estimates of genetic, species, and ecosystem biodiversity, and how do the limitations (i.e., understanding of the scale of error) influence an understanding of biodiversity patterns and of ecosystem structure and function?

Human Impact on Processes Responsible for Biodiversity Change:

1. What are the direct impacts on biodiversity of human-altered systems? That is, what is the variation in biodiversity over spatial and temporal scales relevant to the critical environmental issues? Examples of specific research questions are:

- How do human influences on biodiversity differ from those caused by natural processes?
- To what extent do human effects alter the probability of ecosystem collapse in different systems?
- To what extent are particular changes in biodiversity due to human activities reversible?
- Given the often direct impacts on certain target species within a region, are species within functional groups interchangeable within a system?
- How does the addition or loss of species due to human activities affect community structure and resilience?

2. What are the indirect impacts on biodiversity of human-altered systems? Examples of specific research questions are:

- What characteristics of species enhance susceptibility or provide immunity to precipitous declines?
- In what types of habitats are alternative ecological communities stable?
- Are threshold processes involved in precipitous declines (and the persistence of these declines) in biodiversity, and, ultimately, in the risk of extinction of individual species?
- Does genetic or species diversity provide a buffer against irreversible or massive perturbations?
- What are the long-term effects of species replacements (e.g., exotic species) on ecosystem function?

CONTINGENT AND REASONABLE PREDICTION

In this proposed research program, where a major goal is prediction of the effects of human activities on marine biodiversity as input to conservation and management decisions, it is important to recognize and characterize the *uncertainty* associated with such predictions. Prediction stems from a basic understanding of system processes, and this initiative is designed toward this end. Moreover, this initiative emphasizes the need to expand the predictive capability of the marine sciences within a framework contingent on appropriate spatial and temporal limitations, as well as on limitations in other aspects of the science. Such limitations should be expressly considered in developing predictions. Mechanistic models are valuable tools for predicting and extrapolating beyond the specific parameter range studied, and sensitivity analyses for terms in the models can provide at least qualitative estimates of the uncertainty associated with such predictions and extrapolations.

In the ocean, no diversity inventory can ever be complete, and due to the vastness of even regional sites, large patterns must be discerned with proportionately few samples. Therefore, *inference* must be an important tool in the study of marine biodiversity. Thus, in order to assure the efficiency of research and the rigor of predictions, it will be necessary to understand fully, to develop, and to advance strong *inferential* methods (e.g., Platt, 1964).

In addition, the consequences of taxonomic error are not typically addressed in ecological studies, but appreciation of such consequences is critical to evaluating the reliability of the biodiversity patterns that would form the foundation of the research conducted in this program. Thus, the taxonomic component of this program includes formal evaluation of the consequences of taxonomic error and changing taxonomic resolution on the design, analysis, and inferences drawn from the biodiversity studies (Carney, 1993; Carney, in press).

APPROACHES

Theory and Modeling

The central biodiversity research questions require the means to describe and characterize patterns of biodiversity, an understanding of biodiversity for ecosystem processes, and an understanding of the mechanisms that maintain biodiversity. Each of these challenges involves relating patterns and processes across multiple scales of space, time, and organizational complexity and mandates the use of modern approaches to modeling and computation. Which details at fine scales are important to understanding patterns on broader scales, and which are irrelevant? How can the dynamics of aggregations—populations, communities, or other groupings—be understood in terms of the collective motions of the individuals? How much functional redundancy exists in ecosystems, and how much functional detail is needed to account for shifts in ecosystem response?

Each of these questions involves a recognition that systems are structured into components that have characteristic spatial and temporal scales, and that only macroscopic descriptions of the dynamics within components may be relevant to the dynamics of the *collections* of components. But to determine the identity of those macroscopic descriptors requires a melding of bottom-up, individual-based approaches and top-down phenomenological approaches. Only in this way can the mechanistic basis of the patterns observed be explored, for example, in remote sensing and other broad-scale approaches to description. This also enables the broad-scale predictions of climate models to be interfaced with the much finer-scale understanding that exists at the level of the individual.

The primary frameworks for describing the dynamic movements of populations and communities are those associated with the concepts of Lagrangian and Eulerian motion. In the former, individuals (or packets of fluid) are traced as they make their way across a landscape or seascape. In the latter, particular regions of space provide the starting point, and numbers of individuals (or the densities of materials) in the region are tracked, describing changes in terms of fluxes; individual identities are suppressed. The two approaches provide different perspectives on the dynamics of population and communities.

The study of biodiversity must utilize both modeling approaches. Individual responses to changing environmental cues, and to each other, embody knowledge of organismal-level behavior and physiology and allow the representation of taxonomically based functional biodiversity. But observations are most easily expressed in Eulerian terms, reflecting patterns, for example, of the type detected by remote sensing. Hence, exciting and powerful new advances in modeling will involve the development of individual-based models that allow prediction at the ensemble (aggregate) level, or in an Eulerian frame of reference, and techniques for relating Lagrangian and Eulerian models (e.g., Grunbaum, 1992; Grunbaum and Okubo, 1994).

Great advances have been made in numerical modeling of the oceans in the past few decades. Models can be used to predict the variability in physical properties of the seawater, the general circulation of the oceans, and small-scale changes in flow patterns that might result from anthropogenic causes (such as the building of groins, etc.). Of particular concern within the context of this initiative is the incorporation of biology into physical models. Until recently, organisms have been treated as passive (i.e., nonswimming, noninteractive) tracers in the models. Ideally, dispersal and recruitment can be predicted, given a larval source. In addition to all the usual uncertainties associated with numerical modeling of geophysical fluid dynamics, some of the physical limitations of these models that are particularly relevant to this program include the uncertainty in the diffusivity coefficients and the poor resolution of vertical motions, especially on short time scales.

Over the last decade or so, much progress has taken place in the development of coupled models of biological, chemical, and physical systems. This has been and is being done both on a global scale using "gross" processes (Fasham et al., 1993) and on smaller scales with more detailed interactions (Hofmann, 1988; Hofmann and Ambler, 1988; Werner et al., 1993; Tremblay et al., 1994). In the coming years, the incorporation of spatial dependence into individual-based modeling (DeAngelis and Gross, 1992) is sure to advance this field significantly.

The utility of numerical models for understanding biodiversity phenomena within an oceanographic setting would be improved considerably by the incorporation of effects such as aggregation or behavioral mechanisms (including swimming and vertical migration) that would tend to clump or separate organisms (e.g., Rothlisberg et al., 1983; Franks, 1992; Eckman et al., 1994). These types of models would be particularly useful in addressing connectivity between sites within a region via larval dispersal and settlement (in the case of planktonic propagules of benthic organisms).

Ecological modelling will also be valuable for designing experiments, interpreting results, and making predictions regarding, for example, the role of life-history characteristics in determining the susceptibility of species to global crashes (leading toward extinction), and the probability of subsequent recovery. In such metapopulation models (Nee and May, 1992), the following four crucial factors are involved in predicting, for a given species, the threshold amount of habitat that can be lost and the distribution of remaining habitat for sustaining the species: (1) the total numbers of semi-isolated populations (patches) of competitively superior and weedy species, (2) the health of those populations (e.g., are they increasing, decreasing, or extinct), (3) the disturbance rate that would obliterate local populations/patches (e.g., epidemics, hurricanes), and (4) the colonization abilities of different species that allow them to locate new patches, which are a function of the species' life-history characteristics and the physical connectivity between different patches.

Such models should be an integral part of this initiative because they explicitly relate the local demographic and life-history characteristics of a species and its most important neighbors, and the physics of their "regional-scale" habitat (see also Mangel, 1993). These models also indicate the clear need for both smaller-scale, local studies of the population biology and ecology of species, and of a larger-scale perspective concerning the physical connectivity between suitable habitat, and thus populations. These models dictate the most critical biological and physical parameters that should be measured or manipulated in the field. Moreover, they can be used to develop management criteria for protected areas, particularly because the model output includes predictions of the minimum number of patches required to sustain and "protect" the species.

Retrospective Analysis: Importance of an Historical Perspective

Recent history provides a valuable yet underutilized guide to the factors controlling the number of species occurring at each of several spatial scales. It also enables estimates of rates of change in species numbers and thus provides an important scale against which to measure future changes (Elmgren, 1984, 1989; Webb and Bartlein, 1992; Ricklefs and Schluter, 1993).

Differences in species richness among geographical regions of the modern ocean are, to a large extent, attributable to differences in the magnitudes of extinction, diversification, and immigration of species during the last 20 million years of Earth history. Studies of geographical range contractions reveal not only which factors might be responsible for local and global extinctions, but more importantly, which characteristics of habitats and regions enable species to persist during crises. Similarly, comparisons of fossil biotas with living ones from the same habitats and regions provide clues about conditions that are favorable to the formation and proliferation of species. Such comparisons are especially meaningful and powerful when they are done in conjunction with phylogenetic analyses, because the latter make it possible to trace ancestor-descendant relationships and provide one means by which to infer when, where, and how frequently new species arise. Finally, the fossil record offers both a uniquely long-term perspective on biotic interchange—the movement of species across geographic barriers that have partially or completely broken down (Vermeij, 1987, 1991a, 1991b)—and on the temporal pattern of diversity and rates of origin and extinction of marine species (Jackson et al., 1993; Jackson, 1994).

Given that such interchange potentially affects diversity and that it is taking place through human agency at unprecedented rates, it is important to investigate the extent and consequences of historical cases of biotic interchanges at scales ranging from particular habitats to biogeographic provinces. These studies are most profitably done with taxonomically well-characterized groups having a good fossil record. Foraminiferans, mollusks, corals, sea urchins, barnacles, bryozoans, marine birds and mammals, diatoms, and calcareous algae would thus be particularly promising groups for study.

Another promising historical approach that has been almost entirely ignored is the systematic study and evaluation of museum or archived collections (Box 10) and of monographs from the eighteenth through the early twentieth centuries. Buried in these early collections and papers are important records of species whose ranges have contracted or expanded since then, or which have been wholly exterminated. An example is the logbooks of whalers in the nineteenth century, whose observations of whale populations form the basis of estimates of pre-exploitation population sizes (Evans, 1987; Baker and Palumbi, 1994) and whose pages may literally provide estimates of genetic diversity in previous centuries. From such material it may be possible to estimate rates and times to extinction, rates and times of species introductions and invasions, and other changes in geographic distributions of common species during the last two centuries. This historical research would profitably include in-depth examination of documents related to shipping and fisheries, as well as the proceedings of local scientific societies and of groups devoted to natural history.

Box 10: Museum collections of biological specimens are an invaluable and largely underutilized resource for biodiversity studies.

THE UTILITY OF MUSEUM COLLECTIONS

Some of the first attempts to describe the diversity of species at the base of the food web were made by fisheries biologists in the 1870s and 1880s. These collections, made by biologists aboard the U.S. Fisheries Steamers *Fish Hawk, Albatross,* and *Blake,* can be found today on shelves deep in the recesses of the Smithsonian Institution's National Museum of Natural History. Many tens of millions of other specimens from the waters of the United States and around the world are preserved in that and other museums in this country.

These specimens represent a partial catalogue of the Earth's natural history, and, for those species that are described, provide the standards for names of the world's species. But these specimens can also be used to document biotic changes in areas of habitat alteration, to analyze long-term changes in species distributions, and to determine both the historical and modern importance of the incorporation of human pollutants into marine food webs. Thus, museum specimens of tuna (*Thunnus* spp.) were analyzed to determine the historical trends of mercury contamination.

In addition, museum collections provide a major potential resource for determining species extinctions of both animals and plants in the ocean. This approach could include: (1) locating specimens of species (especially from well-collected coastal habitats) that have not been found again in the twentieth century, and (2) re-examining, through molecular genetic analysis, preserved specimens of extinct populations of species that have been thought to be conspecific with extant populations.

Key References: Lee et al. (1978); Carlton (1993); SA2000 (1994).

Archived, long time-series data are also invaluable for retrospective studies. Long-term sets of physical, chemical, and biological data exist that can provide historical records up to about one century old. The more than five decades of data from the California Cooperative Ocean Fisheries Investigations (CalCOFI) program is one example (Chelton, 1983). Historical accounts and museum collections (Box 10; Jackson, 1994) provide resolution extending to several centuries. Archived cores from individual corals, analogous to tree ring cores, can yield valuable historical information on reef history, El Niño-Southern Oscillation events, and other climatic episodes, for up to perhaps five centuries (Glynn and Colgan, 1992; Dunbar and Cole, 1993). Finally, archived sediment cores and cores of entire coral reefs provide historical environmental, biological, and geological data ranging from thousands of years (with resolution at the level of decades) to hundreds of thousands or several million years (with resolution ranging from thousands to tens of thousands of years) (Jackson, 1992; Webb and

Bartlein, 1992; Valentine and Jablonski, 1993). The use of such existing samples or data sets will allow examination of temporal and spatial scales precluded by short-lived grants, and may also prove to be less expensive than new field studies.

In summary, the historical approaches outlined here offer a perspective on marine biological diversity at longer time scales. They place taxonomic, ecological, and biogeographical studies in a larger framework in which the appearance and disappearance of species can be linked to known mechanisms and events of environmental change.

Linking Biotic Surveys with Ecological and Oceanographic Experiments

Within a regional system, taxonomic surveys of those taxa fundamental to pattern-level and process-level questions should be done as an integral part of the research program. In this sense, biotic surveys are vitally linked to the ecological and oceanographic regional-scale approach identified here.

Many biotic surveys currently exist through fisheries or agency-based activities. Fishery research vessel cruises, for example, sample the ocean widely but typically target a relatively narrow range of species; with additional effort these surveys could be expanded to include more comprehensive biological sampling. Similarly, sampling from fishing vessels, many of which have scientific observers on board, would be an additional source of information.

Marine fisheries also represent one of the greatest manipulations of marine ecosystems by the human race. Management approaches take into account not only biological, but also social, economic, and international considerations. Although not widely applied, the approach of adaptive fisheries management, where fish populations are manipulated to learn about the processes regulating their population sizes (Walters and Hilborn, 1978; Collie, 1991), would provide excellent opportunities for studying attendant biodiversity-related issues within the context of this initiative. In a similar manner, some research under this initiative should take advantage of fishery management regimes to examine ecosystem response carefully and to monitor, and ultimately predict, the consequences in terms of biodiversity.

METHODS

Studying the effects of environmental change and anthropogenic activities on regional-scale marine biodiversity and the consequences to ecosystem function presents unique conceptual and methodological challenges. Much previous ecological work has focused on relatively circumscribed spatial/temporal scales that do not adequately account for the vast geographic ranges of many species. Conversely, relevant spatial scales for smaller organisms and microorganisms have often been inadequately described and undersampled. In addition, although very good examples of environmentally relevant taxonomic work exist, detailed taxonomic studies have been conducted on very limited numbers of representative taxa. Often, there is little or no communication between different taxonomic specialists working in the same habitat or between taxonomists and ecologists working on similar systems. Conceptually then, a wider purview of sampling scales, taxonomic groups, and improved integration of taxonomic and ecological disciplines is necessary for understanding the nature, function, and stability of biodiversity in the marine environment.

Taxonomic Methods

A fundamental underpinning of any biodiversity research is accurate recognition and classification of intraspecific variation, species, and assemblages. Such information is requisite for identifying patterns of biodiversity and for understanding how these patterns change over time and space. Currently, even simple identification of species in many taxonomic groups is severely limited. One reason for this deficit is that traditional taxonomy, and taxonomists themselves, have become "rare and endangered" species for the various reasons discussed earlier. In addition, a lack of convenient, practical, or appropriate methods has hampered rigorous taxonomic approaches in many ecological studies. Therefore, one critical component of an environmentally relevant marine biodiversity research program is the development and application of appropriate methods for identifying and enumerating species, their intraspecific variation, and patterns of biodiversity. Integration of methods with realistic sampling strategies over relevant spatial scales is also a critical component of this program.

Taxonomists are often relied on too heavily for routine kinds of identification work. Some of this load can be reduced by having ecologists better trained in the fundamentals of taxonomy, including the principles associated with taxonomic description (alpha taxonomy), and by training "parataxonomists," persons expert in the identification of the flora and fauna of a particular region. Sufficiently high levels of quality could be maintained through the use of regional centers of taxonomic expertise, where ecologists could send voucher specimens for verification or that could be visited for workshops. At present there is no formal structure for checking identifications made during an ecological study. In contrast, when funding is granted from the Environmental Protection Agency (EPA) for studying anthropogenic effects (e.g., clam raking, dredging or trawl marks, etc.) on a benthic soft-sediment community, for example, there is a requirement to submit a quality assurance/quality control plan which documents the fact that some percentage of sample replicates will be sent blind to another facility for chemical analysis. In order to maintain funding from the EPA, the analyses conducted in a researcher's lab must match closely the results from those of the other facility. The simple deposition of voucher specimens for ecological studies is inadequate, since most of the material already deposited in the U.S. National Museum is reviewed rarely, if ever. Regional centers would help to ensure examination of voucher material and correction of errors.

The ability of ecologists to make accurate species identifications can be improved by the production of readily accessible identification aides. Most keys to marine invertebrates that are currently used, for example, were written in the 1960s or early 1970s and generally use detailed and often arcane terminology so that a user cannot identify most species treated without extensive prior knowledge or training. Moreover, with a few exceptions (e.g., the South Carolina fauna by Ruppert and Fox [1988], the Carolina flora by Schneider and Searles [1991] and the Bermuda fauna and flora [Sterrer, 1986]), keys to marine invertebrates and plants are being published as government documents, marine laboratory reports, or small museum contributions. At the very least, guides to regional faunas and floras, written and illustrated in ways that bring attention to similarities of related taxa, on the model of popular bird guides, should be the first step. Recent software advances make it possible to produce computer-based, iterative, possibly expert-system based or polythetic keys tied to images (still and moving) stored on CD-ROM (Estep et al., 1989, 1993; Pankhurst, 1991; Schalk, 1994).

Other means of disseminating taxonomically important information should also be considered. For example, regional-scale databases containing records of occurrences of, for example, marine algal and invertebrate species do not exist in any form other than simple species lists. Local collections of preserved specimens need to be maintained and augmented. On a national level, facilities such as protist (microalgae and protozoa) culture collections may need augmentation for those groups of living things that cannot be preserved.

Molecular Genetics

Molecular techniques are providing a suite of methods for quantifying the phylogenetic relatedness between species and higher taxa, for recognizing intraspecific genetic variability, and for quantifying difficult-to-identify species and larval forms in natural samples (Box 11). Molecular approaches may provide quantitative resolution not achievable using morphometrics and other traditional taxonomic criteria.

Biotechnology, along with instrumentation innovation, have rapidly accelerated data acquisition and analysis and analytical capability for molecular biodiversity research. Use of oligonucleotide (Baker et al., 1990; Bucklin et al., 1992; R.R. Olson et al., 1991) and immunological (Miller et al., 1991; Demers et al., 1993) markers for identifying intraspecific variability and species are becoming more standard approaches (Yentsch et al., 1988; DeLong et al., 1989; Ward, 1990; Giovannoni and Cary, 1993). Development of the polymerase chain reaction (PCR) for rapid access and replication of individual genes and genetic loci, in combination with improvements in automated DNA sequencing, have greatly expanded the range of possibilities. Combining PCR with high through-put,

---

Box 11: The expanding capability of molecular genetic techniques available for marine science applications will have a large and critical role in addressing many important biodiversity questions.

A CRITICAL ROLE FOR MOLECULAR GENETICS

Molecular genetics has become a basic part of the ecology and taxonomy methodology, and has helped make biodiversity analyses over large spatial scales more practical and more precise. Genetic information, now based largely on DNA analysis, plays two important roles in biodiversity studies—the identification of species and the documentation of gene flow patterns over both small and large spatial scales. Small genetic differences may be used to help identify species whose morphological differentiation is difficult to see or whose morphology is unknown at a critical life-history stage. DNA sequences from marine larvae have been used to discover unexpected species in planktonic assemblages, and similar discoveries have been made for bacterioplankton of the open oceans (see Box 3).

Patterns of genetic variation within species can be used as an indicator of the population history and structure of that species. Should marine species that have global ranges be considered to have one global population? Does overexploitation of a species in one area affect that species elsewhere in the world?

A genetic survey of humpback whales *(Megaptera novaeangliae)*, for example, showed that two populations separated by only 2,500 miles had distinct genetic differences. This is surprising because these animals annually swim more than 8,000 miles in seasonal migrations. These unexpected results indicate that separate populations of this threatened species should be managed as distinct entities to preserve their genetic and population structure. Similarly, divergent evolutionary lines within blue marlin (*Makaira nigricans*) have recently been discovered within and between populations in Atlantic versus Pacific Ocean basins. Recognition (and maintenance) of this diversity may have important implications for management strategies.

Key References: Bucklin et al. (1989); Powers et al. (1990); Bucklin (1991); Finnerty and Block (1992); Baker et al. (1993); Alberte et al. (1994); Olsen et al. (1994).

---

automated DNA sequencing allows rapid identification of individuals within populations and of their genetic relationships (Martin et al., 1992; Baker et al., 1993; van Oppen et al., 1994). Thus, long-term regional-scale studies of interspecific and intraspecific genetic variability along current boundaries, in profiles, or separated by wide biogeographic barriers have recently become possible (Avise et al., 1986). Rapid, semi-automated, molecular biological techniques will certainly facilitate identification of species, intraspecific variability, and hard-to-identify taxa and larval forms over the broad regional and temporal scales identified in this initiative.

To a degree, these techniques provide a unified set of methods that can be applied within and across diverse sets of taxa. Molecular methods are now an important component in many biodiversity research programs. Integration and confirmation of such techniques with traditional taxonomic criteria is a necessary and important component of emerging biodiversity research programs worldwide (NERC, 1992).

Instrumentation

In situ field approaches and innovative instrumentation are having large impacts on marine biodiversity research. A good example is the application of flow cytometric techniques in the recent discoveries of phytoplankton diversity (R.J. Olson et al., 1991). Newly recognized prochlorophytes in the photosynthetic picoplankton were discovered in oligotrophic ocean gyres (see Figure 6) by recognition of their unique pigment signatures using shipboard flow cytometry (Chisholm et al., 1988). These abundant planktonic organisms were largely unnoticed until application of this new technology. This example illustrates the profit of welding different approaches and technologies. In subsequent molecular studies (Giovanonni et al., 1990; Schmidt et al., 1991), an abundant ribosomal RNA sequence, peripherally related to marine *Synechococcus* sp., was identified in bacterioplankton assemblages from oligotrophic ocean gyres. Subsequent genetic analysis of cultured marine prochlorophytes showed that this sequence was derived from the very same prochlorophytes previously identified by flow cytometry.

Use of tagged molecular markers such as fluor-labeled oligonucleotides or antibodies, in combination with sensitive instruments such as flow cytometers, have great potential for assays of biodiversity in natural and anthropogenically altered systems (Ward and Carlucci, 1985; DeLong et al., 1989; Amman et al., 1990).

Sampling

Improvements in sampling methods and in situ approaches will be useful for regional-scale studies of biodiversity (Box 12). For pelagic studies, in situ plankton samplers and fluorimeter devices are being developed for use in tandem with sensitive molecular identification methods for detecting and quantifying the presence and variability of specific larval and microbial taxa. Real-time surveys of variability and abundance of microbial and metazoan organisms are now entering the realm of possibility by coupling sophisticated electronics with molecular identification techniques, but more development will be needed for widescale application of such technology. Ribosomal RNA probes already have been used in conjunction with flow cytometry in analyzing mixed microbial populations (Amman et al., 1990). In situ field sampling and detection represents an important, achievable advance in broad-scale marine biodiversity studies. Remote

Box 12: Conducting regional-scale research will require the application of many new sampling and observing technologies.

THE ROLE OF NEW SAMPLING AND OBSERVING TECHNOLOGIES

Techniques long used in terrestrial studies of biodiversity are now being extended to marine systems. Remote-sensing technology provides a critical tool to evaluate linkages between ecological and oceanographic processes in regional-scale studies. Satellite information can now be used to determine primary production and physical transport processes that influence biodiversity. Navigational capabilities are now accurate to one centimeter, and geographical information systems (GIS) allow storage of images and data at a variety of spatial scales. These advances revolutionize the ability to analyze long-term biological and physical-oceanographic data on many temporal and spatial scales in conjunction, for example, with data on climate changes. In turn, marine scientists must be closely involved in the design and application of satellite sensors.

In sublittoral habitats, obtaining information for the study of biodiversity has been more difficult. Now, however, 10-meter contour topographic maps are available for some sites and centimeter-scale maps are possible. Bottom images from high-resolution side-scan sonar can provide an intermediate-scale reference so that photographic images can be overlaid precisely on multibeam sonar topographic maps. These recently developed techniques can begin to provide a spatial information reference taken for granted in terrestrial studies. Low-cost remotely operated or autonomous vehicles can be launched from shore stations to study the horizontal extent of events observed either from satellite imagery or in situ observing systems. In order to study adequately processes controlling biodiversity, continuous data in both space and time are needed to record low-frequency events. The resultant maps and time-series are needed to guide sampling and to design experiments.

Developments in the assimilation of data using nested series of models will enable analysis of the vast amounts of spatially continuous, real-time data generated by ocean-observing systems (in moorings, bottom landers, and via electro-optical cable connections to land). Awareness of the ways in which different organisms interact with oceanic processes is only just beginning. There are new optical sensors for every scale of resolution from satellite-captured views from space to in situ microscopes; acoustic sensors developed for detecting submarines can be used to study whales, and other acoustic instruments are being used to study fish and their microscopic food. Moreover, it is now possible to visualize the physical, chemical, and biological surroundings of individual planktonic larvae or zooplankton.

These first-hand views of biological as well as physical patterns and structure in the ocean will lead to better scientific questions, better experiments, and better management of marine systems.

Key References: Greene et al., (1988); Currin et al., (1990); Greene and Wiebe (1990); Abbott and Chelton (1991); GLOBEC (1991, 1992); Von Alt and J.F. Grassle (1992); USGS (1994).

sensing and acoustic techniques are also important methods for providing valuable information on regional-scale phenomena (Pieper and Van Holliday, 1984; Greene et al., 1989; P.E. Smith et al., 1989; Fua et al., 1990; Abbott and Chelton, 1991). Although these techniques may not have sufficient resolution to provide specific information on species or intraspecific biodiversity, they can provide broad-scale perspective on bulk environmental properties or processes that may influence, or be affected by, changing patterns in biodiversity. Moreover, high spectral resolution imagery, for example, may be able to distinguish different phytoplankton pigment groups using satellite-based sensors. Carder et al. (1993) used an aircraft-based sensor to study the distribution of substances other than chlorophyll in the coastal ocean. Broad-scale changes in pigment composition may be indicative of changes in the physical environment (Letelier et al., 1993).

New, automated, direct-sampling technology, such as the Moored, Automated, Serial Zooplankton Pump (MASZP) (Doherty and Butman, 1990; Butman, 1994) and new optical imaging technologies such as the Video Plankton Recorder (VPR) for zooplankton (Davis et al., 1992), the Rapid Sampling Vertical Profiler (RSVP) for phytoplankton (Cowles and Desiderio, 1993), and the Eco-Scope (Kills, 1992) for studying predator-prey interactions show great promise for providing species-specific distributions of planktonic organisms over relatively large geographic regions. Some of the best opportunities for technological advances in biological oceanography involve coupling automated, direct-sampling methods with automated, sample-processing techniques. Work is in progress (C.A. Butman and E.D. Garland, pers. comm., 1994), for example, to develop species-specific, fluor-tagged immunological markers for planktonic larvae of benthic invertebrates that can be applied to samples collected by the MASZP. Then, the number of fluorescing organisms can be quantified automatically using digital, color image-analysis techniques (Bjørnsen, 1986; Amman et al., 1990; Berman, 1990; Sieracki and Viles, 1990). This is a good example of the enriching opportunities to *couple* existing technologies developed in diverse fields of science and engineering for new applications in biological oceanography and marine ecology.

Special Opportunities

Coordination with "special opportunities" in environmentally relevant marine biodiversity research should be recognized. In particular, recognition of threatened species in fisheries or inventory studies, or identification of invasive species may provide unique "samples of opportunity" in marine biodiversity research. One example is coordinated studies in collaboration with fisheries experts on recently impacted or endangered species or populations. Other coordinated studies on biodiversity could readily take advantage of long-term, process-oriented biogeochemical time-series studies already in place. Such opportunities should not substitute or supplant, but rather augment, the organism-oriented biodiversity emphasis of this initiative. Finally, the establishment of marine reserves offers one of the most striking opportunities for understanding human impacts in the sea (Box 13).

It is clear that understanding of the distribution of marine organisms, and therefore their conservation and management, will require studies at unprecedented geographic scales. The marine laboratories of the world have great potential to provide the infrastructure and focus for programs in research, training, and education, and the conservation of marine biodiversity. Regional marine laboratories encompass the geographic scale of environmental and ecological gradients and bridge the disciplines of oceanography and ecology, and their region-wide data sets are fundamental to structuring comparative studies of marine biodiversity (Lasserre et al., 1994).

Marine laboratories are found in virtually every coastal country, often in relatively undisturbed locations, with ready access to many representative coastal habitats and organisms. The great majority of laboratories are tied to academic institutions or museums, with long-standing traditions in the study of marine organisms, training of scientists and managers, communication and exchange with other laboratories, and environmental impact assessment. Many marine laboratories are government-supported, with strong mandates for resource management. Their continuity of research and management sets marine laboratories apart from other institutions. They either possess or have direct access to unique, long-term data sets that form a critical baseline against which human impact may be assessed and interpreted.

Whereas marine laboratories are found within different countries or regions with different cultures, they have a common scientific culture and traditions which predispose them to cooperative programs and to networking. For example, the 27 laboratories of the Association of Marine Laboratories of the Caribbean (AMLC) have held annual meetings for almost 30 years. In 1990, with the support of the National Science Foundation and private foundations, over 20 Caribbean laboratories formed the CARICOMP (Caribbean Coastal Marine Productivity) network to conduct comparative, standardized observations of coastal ecosystem structure and function (Ogden, 1987). More recently, 80 European laboratories have joined together in the Marine Research Stations Network (MARS), and U.S. marine laboratories have formed the National Association of Marine Laboratories (NAML), as well as regional groups such as the 35-member Southern Association of Marine Laboratories (SAML) (Lasserre et al., 1994).

IMPLEMENTATION

In order for this national research agenda to be realized, the scientific community, federal agencies, and key related programs will need to work together in a coordinated, committed fashion, tapping into and building on the mounting enthusiasm for tackling the challenges of the critical environmental issues now facing the oceans. A consequence of this realization will be that the "payoffs"

Box 13: Marine research reserves provide many special opportunities for addressing biodiversity research questions in controlled and protected settings.

MARINE BIODIVERSITY RESERVES FOR RESEARCH, CONSERVATION, AND MANAGEMENT

Marine reserves are essential for measuring human impacts on the diversity and stability of communities and for developing more scientifically rigorous guidelines for their protection and management. Whereas manipulative experiments to alter the abundance of selected species or to evaluate environmental changes under controlled conditions are the best way to address many important biodiversity questions, experiments have rarely been conducted at the large scales appropriate to understanding why communities change. Marine reserves would be excellent for addressing questions such as the following:

• What are the physical effects of human activities on coastal biodiversity? Extraordinary effects were demonstrated, for example, when people were excluded from rocky intertidal shores of Chile, resulting in nearly total transformations of intertidal communities in only a few years.
• How large an area is required to protect nursery populations that are fished elsewhere, and over how large an area might such nurseries be effective? Chilean workers have demonstrated that some marine reserves do not work well in supplying adjacent regions with larvae of overharvested species.
• How well can degraded environments and communities be restored, and how long does it take? How effective is the re-establishment of endangered species into reserves? Sea otters, for example, have been re-established successfully in their native habitat in the Pacific Northwest.

There are many excellent opportunities to conduct large-scale human exclusion experiments within existing parks, sanctuaries, military bases, and other areas where public access is already restricted. Areas closed due to collapse of fisheries are also good prospects. The number, size, and treatment of reserves need to be carefully planned *in advance* to take advantage of regional knowledge and for statistical analyses. This is critical for experiments investigating the potential for multiple-use reserves. Finally, human-exclusion experiments provide an opportunity for education about important marine resources to facilitate public support for hard management decisions, particularly those involving recreational activities. Support for such restrictive measures could be increased by involving local residents in experiments.

Key References: VanBlaricom and Estes (1987); Duran and Castilla (1989); D'Elia et al. (1991); Siegfried (1994).

and products that can result from this first national marine biodiversity agenda will contribute significantly to the nation's emerging agenda on the conservation and wise use of the nation's biological resources. This initiative will require major new funding over a decadal time scale to achieve the objectives outlined in this report. In addition, existing programs and resources from federal, state, and private sources would augment any new funding (e.g., see Box 13).

We address here the broad aspects of implementation and directions for this marine biodiversity initiative. Future study plans will develop the specific processes and mechanisms for accomplishing the objectives of this initiative.

This proposed initiative would be propelled by the interactions and involvement of several agencies working under coordinated research umbrellas. Energetic and intimate working relationships with national and international biodiversity programs, marine laboratory networks, museums, and newly emerging fields (e.g., Box 14) will form critical research bridges. This initiative is envisioned in terms of a decadal time scale—a minimum time period to undertake the research efforts proposed here and to achieve a critical level of coordination.

The research questions posed in this agenda are envisioned as being addressed by both small and large research groups seeking designated funds from several agencies through a peer-review process. Once launched, early steps would include regional integration of research efforts. Coordination may include specific efforts to use the same experimental techniques and sampling methods to address the same or similar questions across different systems. The selection of which systems to study in which geographic regions should ultimately be determined by the competitive proposal process. Important decision criteria for funding would include the extent to which the proposed research addressed scientifically perceived environmental threats to the identified study system and the likelihood of achieving substantial new insights that can be applied to conservation and management. The regional-model system approach has the virtue of concentrating effort, but could risk too great a focus on special cases that may be so exceptional as to limit future applications and generalizations. To minimize such biases, regional-model systems should be chosen as much for their diversity of characteristics as for their taxonomic variety and imperilment.

This marine biodiversity initiative is not calling solely for entirely new data collections, experiments, and investigations. The existence of critical, and especially, long-term information or archived samples for a given site, region, taxon, and so forth, will be invaluable for interpreting results of new studies undertaken as part of this program. Such data would decidedly leverage time, effort, and resources spent on conducting new studies. In some special cases, existing data sets (for example, the CalCOFI data mentioned earlier) may be sufficient for addressing a given suite of biodiversity research questions. As noted by Angel (1992), "The information base on which to develop a predictive understanding of the interaction between diversity and ecological process can be greatly enhanced relatively inexpensively by systematically collating existing data and working up extant collections of material."

Box 14: *Knowledge resulting from this research program should have important economic impacts, for example in the newly emerging enterprise of marine biotechnology.*

MARINE BIOTECHNOLOGY AND MARINE BIODIVERSITY: ARE WE LOSING RESOURCES BEFORE THEY ARE EVEN RECOGNIZED?

The newly emerging and rapidly developing field of marine biotechnology depends for its very existence on a constantly new and broadening knowledge of resource organisms for biomedical, bioengineering, and mariculture research. The potential application of information that could be derived from this initiative for the biotechnological frontier includes natural products derived from marine animals and plants in all ocean habitats, preservation of marine genomic information, development of hardy culture stocks, resistance to disease, and closed-system, computer-controlled aquaculture of marine species. These are but a few examples of the total potential waiting to be tapped.

Loss of biodiversity in tropical terrestrial ecosystems, long before it is even recorded, and the concomitant loss to the biomedical and biotechnological sciences, are now well documented.

Loss of biodiversity in the world's oceans (especially given the much greater higher-order diversity in the oceans, and thus the potential loss of the genome and chemical makeup of an entire class or order of organisms), and its potential value for solving pressing medical and food problems of the growing human population would be one of the great tragedies of our time.

Key References: Colwell (1983); Fautin (1988); Marine Life Resources Workshop (1989); Colwell and Hill (1992); Weber (1993).

RELATIONSHIP TO OTHER PROGRAMS

"An apothegm applies here: a person with one watch knows what time it is; a person with two watches is never sure. Those who espouse solely the approach based on carbon stocks and flows, or solely the approach based on food webs, can confidently contrast marine and terrestrial biotas. Those who consider both approaches have grounds for further thought."

Cohen (1994, p. 63)

This initiative proposes research on regional scales not previously undertaken in the marine environment. The decision to work at these larger scales suggests fundamental and potential linkages and synergisms with other programs that have similar or overlapping interests. In addition, the missions of many federal agencies increasingly call for addressing critical marine biodiversity issues. This initiative provides the next step to connect and coordinate research opportunities with other programs and with agency interests and efforts. In turn, oceanographic-ecological coupling within a regional-model system approach means that this initiative could provide numerous potential benefits for other biodiversity programs.

Scientific Programs and Initiatives

The Systematics Agenda 2000 (SA2000, 1994) is a consortium-based program that seeks to improve the discipline of systematics and taxonomy throughout the U.S. scientific structure. The importance of accurate and reliable taxonomy to studies of biodiversity cannot be overemphasized. Thus, the goals of SA2000 coincide with some of the goals of this initiative—enhancing taxonomy as a discipline and improving the scientific knowledge base of the systematics of marine organisms. This marine biodiversity initiative further recommends an enhanced interrelationship between taxonomists and ecologists. Systematic support groups, such as the Association of Systematics Collections (ASC), have formulated plans for the conservation, support, and use of the nation's museum collections.

The Diversitas program, sponsored by the International Union of Biological Sciences (IUBS), the Scientific Committee on Problems of the Environment (SCOPE), the International Union of Microbiological Scientists (IUMS), and the United Nations Educational, Scientific and Cultural Organization (UNESCO) is a broadly based program in biodiversity covering all of the Earth's ecosystems (diCastri and Younès, 1990; 1994). Under the auspices of the International Council of Scientific Unions (ICSU), Diversitas has become one of the four components (along with the World Climate Research Program, the International Geosphere-Biosphere Program, and the Human Dimensions of Global Environmental Change Program) of a comprehensive program of Earth systems research (Perry, 1993). The scientific questions asked by Diversitas are closely related to those in this initiative. The marine component of Diversitas is described by J.F. Grassle et al. (1991); the European MARS Network, noted earlier, is a component of Diversitas. Diversitas has sponsored workshops on biodiversity and ecosystem function for many habitats, including coral reefs; upwelling systems; estuaries, lagoons, and nearshore coastal systems; and pelagic systems. These workshop reports and conclusions are a rich source of research questions closely related to those identified here.

This marine biodiversity initiative would benefit—and would benefit from— some of the other major marine initiatives such as the Joint Global Ocean Flux Study (JGOFS) and the Global Ocean Ecosystem Dynamics (GLOBEC) program. One of the objectives of JGOFS is to "understand the global-scale processes that control carbon, nitrogen, oxygen, phosphorus, and sulfur exchanges in the ocean over time" (NRC, 1994b). The goal of GLOBEC is "to predict the effects of changes in the global environment on the abundance, variation in abundance, and production of marine animals," with an emphasis on how changing climate alters the ocean's physical environment and how such alterations affect zooplankton and fish (NRC, 1994b).

This proposed initiative has clear synergistic linkages with programs like JGOFS and GLOBEC; research questions called for here provide a framework, for example, for addressing the role of biodiversity (particularly at the microbial level) in ocean carbon flux, and thus the precise biological pathways through which carbon may pass in the ocean. Similarly, this initiative identifies global climate change as a critical environmental issue relative to potential changes to the oceans' biodiversity, and thus this initiative forms a natural linkage with other efforts that seek to understand the biological and oceanographic mechanisms that may influence the temporal and spatial variation of populations in the sea.

Federal Agency Programs

Numerous federal programs address a wide variety of biodiversity technologies and issues. Programs within the Departments of Commerce, Interior, Energy, and Defense, the National Institutes of Health (NIH), the National Science Foundation (NSF), and the National Science and Technology Council (NSTC), focusing on such aspects as genome mapping, new information technologies, bioremediation, and environmental quality assessment, offer important opportunities for interaction with many aspects of this proposed marine biodiversity initiative, as well as for the cooperative use of resources.

Within NSF, research on marine biological diversity is funded within the Ocean Sciences Division (Directorate for Geosciences), the Office of Polar Programs, and the Biological Sciences Directorate. The first two offices currently focus primarily on basic science issues relating to: ecological, evolutionary, and historical processes responsible for maintaining or changing diversity in a system; the functional role of diversity in ecosystems and populations, and the importance of species/gene redundancy; the impact of introduced species on natural ecosystems; developing advanced procedures for resolving taxa in traditionally recalcitrant groups (e.g., microbes, algae, planktonic invertebrates); characterization of diversity in marine systems, including coastal, open-ocean, sea ice, and deep-sea systems; and biogeography related to dispersal processes. These offices are also concerned about the declining expertise in taxonomy and systematics, and the ongoing maintenance of living collections and laboratory facilities for research in marine biological diversity.

NSF's Biological Sciences Directorate (primarily through the Division of Environmental Biology, but including the Divisions of Molecular and Cellular Biosciences, and of Integrative Biology and Neurosciences) currently supports some marine biodiversity research activities as part of its general funding of biodiversity, including Biotic Survey and Inventory Research, Basic Research in

Conservation and Restoration Biology, Partnerships for Enhancing Expertise in Taxonomy, and the Joint International Research program administered cooperatively with the U.S. Agency for International Development (AID).

The National Oceanic and Atmospheric Administration (NOAA), in the Department of Commerce, has a diverse mission that includes promoting environmental stewardship in order to conserve and manage the nation's marine and coastal resources. NOAA has a long history of work in marine biological diversity, stemming from over a century of physical and biological surveys and two decades as the federal trustee for the conservation, protection, and management of fishery resources, marine mammals, and endangered marine species. NOAA is also responsible for managing Marine Sanctuaries and Estuarine Research Reserves, and has the federal mandate to evaluate and restore marine ecosystems impacted by oil spills and other human disturbances.

The 1992 NOAA Strategic Plan (NOAA, 1992) states that the inventory, assessment, and maintenance of marine and coastal biodiversity is an important goal of the element entitled Environmental Stewardship: Coastal Ecosystem Health. New opportunities for large-scale experimental approaches would complement NOAA's efforts to manage and protect fisheries and other species, as well as their habitats. NOAA's National Marine Sanctuary Program and the National Estuarine Research Reserve system have programs to inventory biodiversity and begin the monitoring and assessment of long-term trends within these protected areas. These programs provide yet another cooperative platform for research. As the primary agency concerned with marine biodiversity within the national partnership for biological survey (of which the National Biological Service [NBS], described below, is a part), NOAA's mission and programs align nicely with the research framework provided by this marine biodiversity initiative.

The Office of Naval Research (ONR), within the Department of Defense, relies on continual improvements in the understanding of marine biodiversity and its impacts on natural marine communities. As a steward of the marine environment, ONR programs target research that contributes to preservation and conservation of marine resources and diversity, and to the development of technologies that eliminate or minimize potentially harmful impacts of naval operations on the marine environment. A major thrust area is the development of biotechnical capabilities that exploit the rich diversity of marine organisms to provide new materials, processes, and capabilities. ONR's programs also target understanding of the ocean biota in relation to the chemistry and physics of the sea to advance the ability to predict ocean dynamics on time and space scales relevant to naval operations.

The Department of Energy's (DOE) Office of Energy Research supports marine biological diversity research through its Ocean Margins Program (OMP) and its Microbial Genome Initiative (MGI). The OMP is concerned with assessing the role of the coastal ocean in affecting climate change and the global carbon cycle. In 1994, DOE launched a new molecular biology initiative within the

OMP. The overall goal of this initiative is to provide mechanistic understanding of complex biological processes which mediate major biogeochemical cycles in marine ecosystems, with primary emphasis on the carbon cycle. Molecular biological techniques are being used to address both carbon reduction and carbon oxidation. These molecular biological techniques will be adapted to, and conducted in concert with, OMP field activities scheduled in the southwestern Middle Atlantic Bight. The initiative will emphasize a high degree of coordination and cooperation between investigators, ranging from molecular biologists to "traditional" biologists, chemists, geochemists, and physical oceanographers.

The MGI, a spin-off from the DOE/NIH Human Genome Program, will develop a microbial genome sequencing capability that will provide genomic sequence and mapping information on microorganisms with environmental or energy relevance, phylogenetic relevance, or potential commercial importance and application. For the first time, scientists will be able to compare, side by side, genomic sequence information from microorganisms with similar physiological attributes and phylogenetic lineage. This information will further understanding and application of marine biological diversity by providing insight into genetic expression and regulation in key marine microorganisms.

Another element in this array of federal activities is the Committee on Environment and Natural Resources (CENR) of the NSTC. The NSTC was established in November 1993 to raise science and technology to the same level of consideration as national security, domestic policy, and the economy. The CENR has been working to develop a national research and development strategy for the federal government on such issues as global change, resource use and management, air quality, toxic substances, natural disasters, and preservation of freshwater and marine environments, as well as to assess relevant social and economic sciences, technologies, and risk assessment techniques. The CENR subcommittees encompass all areas of research on environment and natural resources. One of the CENR's subcommittees, the Subcommittee on Biological Diversity and Ecosystems, is responsible for examining the sustainability of the ecological systems and processes that support life and provide the goods and services necessary for human well-being and opportunity. This subcommittee hopes to develop and promote a research strategy to improve understanding of the interactions among biodiversity, ecosystem dynamics, ecosystem management, and environmental degradation.

Finally, the NBS, a newly established component of the Department of the Interior, has a mission of describing the biological resources of the United States (NRC, 1993). One initial charge is to bring all existing inventory knowledge together at a proposed National Biodiversity Information Center. This national agenda for marine biodiversity research can thus work closely with NBS in achieving its goals, and in particular can provide critical *marine* perspectives as the NBS evolves and grows.

SEVEN

Summary

Propelled by the need for understanding changes in marine biodiversity resulting from human activities, this proposed research program calls for ecological and oceanographic research spanning a broad range of spatial scales, from local to much larger regional, and over appropriately long time scales for capturing the dynamics of the system under study (Box 15). The research agenda proposes a fundamental change in the approach by which biodiversity is measured and studied in the ocean by emphasizing integrated regional-scale research strategies within an environmentally relevant and socially responsible framework. This is now possible because of recent technological and conceptual advances within the ecological, molecular, and oceanographic sciences. A major goal of this research is to improve predictions of the effects of the human population on the diversity of life in the sea, in order to improve conservation and management plans.

A well-defined set of biodiversity research questions is proposed for study in several different types of regional-scale marine ecosystems. These studies will permit meaningful comparisons across different habitats of the causes and consequences of *changes* in biodiversity due to human activities. This agenda requires significant advances in taxonomic expertise for identifying marine organisms and documenting their distributions, in knowledge of local and regional natural patterns of biodiversity, and in understanding of the processes that create and maintain these patterns in space and time. Thus, this program could provide long-awaited, much-needed, and exciting opportunities to develop the interface between taxonomy and ecology and between the ecological and oceanographic sciences.

The ultimate benefit to science and society of this research program (Box 16) would be an enhanced ability for long-term sustained use of the oceans and marine organisms for food, mineral resources, biomedical products, recreation, and other aesthetic and economic gains, while conserving and preserving biodiversity and ecosystem function of life in the sea.

Box 15: *This national marine biodiversity research initiative could be many things to many people.*

THE MARINE BIODIVERSITY INITIATIVE: WHAT IT WOULD BE

- An environmentally responsible and socially relevant basic research program on the causes and consequences of changes in marine biological diversity due to effects of human activities.
- A research agenda guided by well-defined research questions that will be addressed concurrently in several different regional-scale systems.
- A program that focuses on large scales that were previously intractable but are absolutely required to address the most compelling biodiversity research questions.
- A partnership between the ecological and oceanographic sciences, both conceptually and methodologically, for explaining biodiversity patterns, processes, and consequences.
- A partnership between ecology and taxonomy, with a major focus on reinvigorating the field of marine taxonomy and systematics.
- A research program with the ultimate goal of improving predictions regarding future effects of human activities on marine biodiversity, thus facilitating use of the sea for societal needs while minimizing impacts on nature.

Box 16: *This national marine biodiversity research initiative could do many things for many people.*

THE MARINE BIODIVERSITY INITIATIVE: WHAT IT WOULD DO

- Enhance understanding of the fundamental processes responsible for the creation, maintenance, and regulation of marine biodiversity and for changes due to anthropogenic effects.
- Dramatically improve knowledge of the magnitude and distribution of the diversity of animals, plants, and microbes in the marine environment.
- Stimulate the development of new techniques for studying linkages between local (ecological) and regional (oceanographic) processes.
- Stimulate the field of marine taxonomy and systematics, particularly the incorporation of molecular techniques for species identification and population differentiation.
- Provide valuable information for policy makers regarding the preservation and conservation of marine life in the face of rapidly expanding threats from human activities.
- Lead to the long-term, sustained use of the oceans and marine organisms for food, mineral resources, biomedical products, recreation, and other aesthetic and economic gains.

References

Abbott, M.R. and D.B. Chelton 1991. Advances in passive remote sensing of the ocean. U.S. National Report of the International Union of Geologists and Geophysicists 1987-1990, Rev. Geophys. Space Phys. 29:571-589.

Adam, P. 1990. Saltmarsh Ecology. Cambridge University Press, Cambridge, Massachusetts, 461 pp.

Alberte, R.S., G.K. Suba, G. Procaccini, R.C. Zimmerman, and S.R. Fain 1994. Assessment of genetic diversity of seagrass populations using DNA fingerprinting: Implications for population stability and management. Proc. Natl. Acad. Sci. USA 91:1049-1053.

Alpine, A.E. and J.E. Cloern 1992. Trophic interactions and direct physical effects control phytoplankton biomass and production in an estuary. Limnol. Oceanogr. 37:946-955.

Alverson, D.L. 1992. A review of commercial fisheries and the Steller sea lion (*Eumetopias jubatus*): The conflict arena. Rev. Aqua. Sci. 6:203-256.

Amman, R., B.J. Binder, R.J. Olson, S.W. Chisholm, R. Devereux, and D.A. Stahl 1990. Combination of 16S rRNA-targeted oligonucleotide probes with flow cytometry for analyzing mixed microbial populations. Appl. Environ. Microbiol. 56:1919-1925.

Anderson, D.M. 1994. Red tides. Sci. Amer. 271:62-68.

Angel, M.V. 1992. Managing diversity in the oceans. Pp. 23-62 in Diversity of Oceanic Life: An Evaluative Review, M.N.A. Peterson, ed. Center for Strategic and International Studies, Washington, D.C.

Angel, M.V. 1993. Biodiversity of the pelagic ocean. Conser. Biol. 7:760-772.

Avise, J.C., G.S. Helfman, N.C. Saunders, and C.S. Hales 1986. Mitochondrial DNA differentiation in North American eels: Population genetic consequences of an unusual life history pattern. Proc. Natl. Acad. Sci. USA 83:4350-4354.

Backus, R.H. and D.W. Bourne, eds. 1987. Georges Bank, MIT Press, Cambridge, Massachusetts, 593 pp.

Baker, C.S. and S.R. Palumbi 1994. Which whales are hunted? A molecular genetic approach to monitoring whaling. Science 265:1538-1539.

Baker, C.S., S.R. Palumbi, R.H. Lambertsen, M.T. Weinrich, J. Calambokidis, and S.J. O'Brien 1990. Influence of seasonal migration on geographic distribution of mitochondrial DNA haplotypes in humpback whales. Nature 344:238-240.

Baker, C.S., A. Perry, J.L. Bannister, M.T. Weinrich, R.B. Abernethy, J. Calambokidis, J. Lien, R.H. Lambertsen, J. Urban, R.O. Vasquez, P.J. Clapham, A. Alling, U. Arnason, S.J. O'Brien, and S.R. Palumbi 1993. Abundant mitochondrial DNA variation and world-wide population structure in humpback whales. Proc. Natl. Acad. Sci. USA 90:8239-8243.

Bally, R. and C.L. Griffiths 1989. Effects of human trampling on an exposed rocky shore. Inter. J. Ecol. Stud. 34:115-125.

Beatley, T. 1991. Protecting biodiversity in coastal environments: Introduction and overview. Coastal Managmt. 19:1-19.

Beauchamp, K.A. and M.M. Gowing 1982. A quantitative assessment of human trampling effects on a rocky intertidal community. Mar. Environ. Res. 7:279-283.

Behrenfeld, M.J., J.T. Hardy, H. Gucinski, A. Hanneman, H. Lee II and A. Wones 1993a. Effects of Ultraviolet-B radiation on primary production along latitudinal transects in the south Pacific Ocean. Mar. Environ. Res 35:349-363.

Behrenfeld, M.J., J.W. Chapman, J.T. Hardy, and H. Lee II 1993b. Is there a common response to Ultraviolet-B radiation by marine phytoplankton? Mar. Ecol. Prog. Ser. 102:59-68.

Bennett, B.A., C.R. Smith, B. Glaser, and H.L. Maybaum 1994. Faunal community structure of a chemoautotrophic assemblage on whale bones in the deep northeast Pacific Ocean. Mar. Ecol. Prog. Ser. 3:205-223.

Bengston, J.L. and R.M. Laws 1985. Trends in crabeater seal age at maturity: An insight into Antarctic marine interactions. Pp. 670-675 in Antarctic Nutrient Cycles and Food Webs, W.R. Siegfried, P.R. Condy, and R.M. Laws, eds., Springer-Verlag, New York.

Bergh, O., K.Y. Borsheim, G. Bratbak, and M. Heldal 1989. High abundance of viruses found in aquatic environments. Nature 340:467-468.

Berman, M.S. 1990. Application of image analysis in demographic studies of marine zooplankton in large marine ecosystems. Pp. 122-131 in Large Marine Ecosystems. Patterns, Processes, and Yields, K. Sherman, L.M. Alexander, and B.D. Gold, eds. American Association for the Advancement of Science, Washington, D.C.

Billett, D.S.M., R.S. Lampitt, A.L. Rice, and R.F.C. Mantoura 1983. Seasonal sedimentation of phytoplankton to the deep-sea benthos. Nature 302:520-522.

Bjørnsen, P.K. 1986. Automatic determination of bacterioplankton biomass by image analysis. Appl. Environ. Microbiol. 51:1199-1204.

Blake, J.A. and J.F. Grassle. Benthic community structure on the U.S. south Atlantic slope off the Carolinas: Spatial heterogeneity in a current-dominated system. In press, Deep-Sea Res.

Boehlert, G.W. and B.C. Mundy 1988. The role of behavioral and physical factors in fish recruitment to estuarine nursery areas. Amer. Fish. Soc. Symp. 3:51-67.

Boesch, D.F., N.E. Armstrong, C. D'Elia, N.G. Maynard, H.W. Paerl, and S.L. Williams 1993. Deterioration of the Florida Bay Ecosystem: An Evaluation of the Scientific Evidence. Report of the Florida Bay Scientific Review Panel to the Interagency Working Group on Florida Bay and to the Assistant Secretary of the Interior. Frampton, Fla., 27 pp.

Boesch, D.F. and N.N. Rabalais 1987. Long-Term Environmental Effects of Offshore Oil and Gas Development. Elsevier Applied Science, London.

Bothwell, M.L., D.M.J. Sherbot, and C.M. Pollock 1994. Ecosystem response to solar Ultraviolet-B radiation: Influence of trophic-level interactions. Science 265:97-100.

Boudreau, P.R. 1992. Acoustic observations of patterns of aggregation in haddock (*Melanogrammus aeglefinus*) and their significance to production and catch. Can. J. Fish. Aquatic Sci. 49:23-31.

Bowen, B.W., A.B. Meylan, and J.C. Avise 1991. Evolutionary distinctiveness of the endangered Kemp's ridley sea turtle. Nature 352:709-711.

Bowen, B.W., W.S. Nelson, and J.C. Avise, 1993. A molecular phylogeny for marine turtles: Trait mapping, rate assessment, and conservation relevance. Proc. Natl. Acad. Sci. USA 90:5574-5577.

Breen, R. 1994. Personal communication to J.T. Carlton.

Brown, B.E. and J.C. Ogden 1993. Coral bleaching. Sci. Amer. 268:64-70.

Bruton, M.N. and R.E. Stobbs 1991. The ecology and conservation of the coelocanth *Latimeria chalumnae*. Env. Biol. Fish. 32:313-339.

Bucklin, A. 1991. Population genetic responses of the planktonic copepod *Metridia pacifica* to a coastal eddy in the California current. J. Geophys. Res. 96:14799-14808.

Bucklin, A., M.M. Rienecker and C.N.K. Mooers 1989. Genetic tracers of zooplankton transport in coastal filaments off northern California. J. Geophys. Res. 94:8277-8288.

Bucklin, A., B.W. Frost, and T.D. Kocher 1992. DNA sequence variation of the mitochondrial 16S rRNA in *Calanus* (Copepoda: Calanoida): Intraspecific and interspecific patterns. Mol. Mar. Biol. Biotech. 1:397-407.

Burkholder, J.M., E.J. Noga, C.H. Hobbs, and H.B. Glasgow, Jr. 1992. New "phantom" dinoflagellate is the causative agent of major estuarine fish kills. Nature 358:407-410.

Buss, L.W. and E.W. Iverson 1981. A new genus and species of Sphaeromatidae (Crustacea: Isopoda) with experiments and observations on its reproductive biology, interspecific interactions and color polymorphisms. Postilla (Yale University Peabody Museum of Natural History) 184, 23 pp.

Butman, C.A. 1987. Larval settlement of soft-sediment invertebrates: The spatial scales of pattern explained by active habitat selection and the emerging role of hydrodynamical processes. Oceanogr. Mar. Biol. Ann. Rev. 25:113-165.

Butman, C.A. 1994. CoOP, Coastal Ocean Processes Study—Interdisciplinary approach, new technology to determine biological, physical, geological processes affecting larval transport on inner shelf. Sea Tech. 35:44-49.

Butman, C.A. and J.T. Carlton 1993. Biological Diversity in Marine Systems (BioMar), A Proposed National Research Initiative. Report of a workshop sponsored by the National Science Foundation, Washington, D.C., 20 pp.

Butman, C.A. and J.T. Carlton. Marine biological diversity: Some important issues, opportunities and critical research needs. In press, U.S. National Report of the International Union of Geologists and Geophysicists 1991-1994. Rev. Geophys. Space Phys.

Butman, C.A., J.T. Carlton, and S.R. Palumbi. Waiting for a whale: Has human hunting altered deep-sea biodiversity? In press, Conser. Biol.

Butman, C.A. and E.D. Garland. 1994. Personal communication to J.T. Carlton.

Caffey, H.M. 1985. Spatial and temporal variation in settlement and recruitment of intertidal barnacles. Ecol. Monogr. 55:313-332.

Carder, K.L., P. Reinersman, R.F. Chen, F. Muller-Karger, C.O. Davis, and M. Hamilton 1993. AVIRIS calibration and application in coastal oceanic environments. Remote Sens. Environ. 44:205-216.

Carlton, J.T. 1985. Transoceanic and interoceanic dispersal of coastal marine organisms: The biology of ballast water. Oceanogr. Mar. Biol. Ann. Rev. 23:313-371.

Carlton, J.T. 1989. Man's role in changing the face of the ocean: Biological invasions and implications for conservation of near-shore environments. Conserv. Biol. 3:265-273.

Carlton, J.T. 1993. Neoextinctions of marine invertebrates. Amer. Zool. 33:499-509.

Carlton, J.T. and J.B. Geller 1993. Ecological roulette: The global transport of nonindigenous marine organisms. Science 261:78-82.

Carlton, J.T., J.K. Thompson, L.E. Schemel, and F.H. Nichols 1990. Remarkable invasion of San Francisco Bay (California, USA) by the Asian clam *Potamocorbula amurensis*. I. Introduction and dispersal. Mar. Ecol. Prog. Ser. 66:81-94.

Carney, R.S. 1993. Review and reexamination of OCS spatial-temporal variability as determined by MMS studies in the Gulf of Mexico. U.S. Dept. of the Interior Minerals Management Service, OCS study MMS 93-0041, New Orleans, Louisiana, 210 pp.

Carney, R.S. 1994. Personal communication to J.T. Carlton.

Carney, R.S. On the adequacy and improvement of marine benthic pre-impact surveys: Examples from the Gulf of Mexico outer continental shelf. In Environmental Impact Assessment: Conceptual Issues and Applications in Coastal Marine Habitats, R.J. Schmitt and C.W. Osterberg, eds. University of California Press, Los Angeles, in press.

Carroll, L. 1872. Through the Looking-Glass and What Alice Found There. Random House, New York.

Chabreck, R.H. 1988. Coastal marshes. Ecology and wildlife management. University of Minnesota Press, Minneapolis. 138 pp.

Chapman, J.W. 1988. Invasions of the northeast Pacific by Asian and Atlantic gammaridean amphipod crustaceans, including a new species of *Corophium*. J. Crust. Biol. 8:364-382.

Chelton, D.B. 1983. CalCOFI—A 33-year oceanographic survey of the southern California current system. Int. Oceangr. Comm., Tech. Ser. Time Series of Ocean Measurements 1:9-13.

Chisholm, S.W. 1994. Personal communication to C.A. Butman and J.T. Carlton.

Chisholm, S.W., R.J. Olson, E.R. Zettler, R. Goericke, J.B. Waterbury, and N.A. Welschmeyer 1988. A novel free-living prochlorophyte abundant in the oceanic euphotic zone. Nature 334:340-343.

Chisholm, S.W., S.L. Frankel, R. Goericke, R.J. Olson, B. Palenik, J.B. Waterbury, L. West-Johnsrud, and E.R. Zettler 1992. *Prochlorococcus marinus* nov. gen. nov. sp.: An oxyphototrophic marine prokaryote containing divinyl chlorophyll *a* and *b*. Arch. Microbiol. 157:297-300.

Clark, R.B. 1992. Marine Pollution. 3rd ed. Clarendon Press, Oxford, 192 pp.

Cohen, J.E. 1994. Marine and continental food webs: Three paradoxes? Phil. Trans. R. Soc. Lond. B 343:57-69.

Collette, B.B., J.L. Russo, and L.A. Zavala-Camin 1978. *Scomberomorus brasiliensis*, a new species of Spanish mackerel from the western Atlantic. Fish. Bull. 76:273-280.

Collie, J. 1991. Adaptive strategies for management of fisheries resources in large marine ecosystems. Pp. 225-242 in Food Chains, Yields, Models, and Management of Large Marine Ecosystems, K. Sherman, L.M. Alexander, and B.D. Gold, eds. Westview Press, Boulder, Colorado.

Colwell, R.R. 1983. Biotechnology in the marine sciences. Science 222:1329-1331.

Colwell, R.R. and R. Hill 1992. Microbial diversity. Pp. 100-102 in Diversity of Oceanic Life: An Evaluative Review, M.N.A. Peterson, ed. Center for Strategic and International Studies, Washington, D.C.

Connell, J.H. 1961. The influence of interspecific competition and other factors on the distribution of the barnacle *Chthamalus stellatus*. Ecology 42:710-723.

Connell, J.H. 1978. Diversity in tropical rain forests and coral reefs. Science 199A:1302-1310.

Corliss, J.O. 1994. An interim utilitarian ("user friendly") hierarchical classification and characterization of the protists. Acta Protozoologica 33:1-51.

Cowles, T.J. and R.A. Desiderio 1993. Resolution of biological microstructure through *in situ* fluorescence emission spectra. Oceanography 6:105-111.

Currin, C.A., H.W. Pearl, G.K. Suba, and R.S. Alberte 1990. Immunofluorescence detection and characterization of N2-fixing microorganisms from aquatic environments. Limnol. Oceanogr. 35:59-71.

Dahl, T.E., C.E. Johnson, and W.E. Frayer 1991. Status and trends of wetlands in the coterminous United States, mid-1970's to mid-1980's. U.S. Department of the Interior, Fisheries Wildlife Service, Washington, D.C.

Davis, C.S., S.M. Gallager, M.S. Berman, L.R. Haury, and J.R. Strickler 1992. The Video Plankton Recorder (VPR): Design and initial results. Arch. Hydrobiol. Beih. 36:67-81.

Dayton, P.K., B.J. Mordida, and F. Bacon 1994. Polar marine communities. Amer. Zool. 34:90-99.

DeAngelis, D.L. and L.J. Gross, eds. 1992. Individual-based models and approaches in ecology. Populations, communities and ecosystems. Proc. Symposium/Workshop: Knoxville, Tenn., May 16-19, 1990.

deGroot, S.J. 1984. The impact of bottom trawling on benthic fauna of the North Sea. Ocean Managmt. 9:177-190.

D'Elia, C.F., R.W. Buddemeier, and S.V. Smith 1991. Workshop on coral bleaching, coral reef ecosystems and global change: Report of proceedings. Maryland Sea Grant Program, University of Maryland, College Park, Maryland.

DeLong, E.F. 1992. Archaea in coastal marine environments. Proc. Natl. Acad. Sci. USA 89:5685-5689.

DeLong, E.F., D.G. Franks, and A.L. Alldredge 1993. Phylogenetic diversity of aggregate-attached versus free-living marine bacterial assemblages. Limnol. Oceanogr. 38:924-934.

DeLong, E.F., G.S. Wickham, and N.R. Pace 1989. Phylogenetic stains: Ribosomal RNA-based probes for the identification of single cells. Science 243:1360-1363.

DeMartini, E.E., D.M. Ellis, and V.A. Honda 1993. Comparisons of spiny lobster, *Panulirus marginatus*, fecundity, egg size, and spawning frequency before and after exploitation. Fish. Bull. 91:1-7.

Demers, A., Y. Kagadeuc, J.J. Dodson, and R. Lemieux 1993. Immunofluorescence identification of early life history stages of scallops (*Pectinidae*). Mar. Ecol. Prog. Ser. 97:83-89.

Deming, J.W., A.L. Reysenbach, S.A. Macko, and C.R. Smith. The microbial diversity of a whale fall on the seafloor: bone-colonizing mats and animal-associated symbionts. In press, Microsc. Res. Tech.

Denny, M.W. 1988. Biology and the mechanics of the wave-swept environment. Princeton University Press, Princeton, New Jersey, 329 pp.

Denny, M. W. 1993. Air and Water: The Biology and Flow of Life's Media. Princeton University Press, Princeton, New Jersey.

di Castri, F. and T. Younès 1990. Ecosystem function of biological diversity. Biol. Intern. (Special Issue 22), International Union of Biological Sciences, Paris.

di Castri, F. and T. Younès 1994. Diversitas: Yesterday, today and a path towards the future. Biol. Intern. 29:3-23.

Ditullio, G.R., D.A. Hutchins, and K.W. Bruland 1993. Interaction of iron and major nutrients controls phytoplankton growth and species composition in the tropical north Pacific Ocean. Limnol. Oceanogr. 38:495-508.

Doherty, K.W. and C.A. Butman 1990. A time- or event-triggered, automated, serial plankton-pump sampler. Pp. 15-23 in Advanced Engineering Laboratory Projects Summaries—1989, D. Frye, E. Stone and A. Martin, eds. Woods Hole Oceanographic Institution Tech. Rept. 90-20, Woods Hole, Massachusetts.

Duggins, D.O., C.A. Simenstad, and J.A. Estes 1989. Magnification of secondary production by kelp detritus in coastal marine ecosystems. Science 245:170-173.

Dunbar, R.B. and J.E. Cole, eds. 1993. Coral records of ocean-atmosphere variability. Publ. University Corporation for Atmospheric Research (UCAR), 38 pp.

Duran, L.R. and J.C. Castilla 1989. Variation and persistence in the middle rocky intertidal community of central Chile, with and without human harvesting. Mar. Biol. 103:555-562.

Dutch, M. 1988. A characterization of polychaete assemblages on a Hawaiian fringing reef. Master's thesis, University of Hawaii, Department of Zoology.

Ebert, T.A. and M.P. Russell 1988. Latitudinal variation in size structure of the west coast purple sea urchin: A correlation with the headlands. Limnol. Oceanogr. 33:286-294.

Eckman, J.E., F.E. Werner, and T.F. Gross 1994. Modelling some effects of behavior on larval settlement in a turbulent boundary layer. Deep-Sea Res. II 41:185-208.

Ehrlich, P.R. and A.H. Ehrlich 1981. Extinction: The Causes and Consequences of the Disappearance of Species. Random House, New York.

Elmgren, R. 1984. Trophic dynamics in the enclosed, brackish Baltic Sea. Rapp. P.-v. Reun. Cons. Int. Explor. Mer. 183:152-169.

Elmgren, R. 1989. Man's impact on the ecosystem of the Baltic Sea: Energy flows today and at the turn of the century. Ambio 18:326-332.

Estep, K.W., A. Hasle, L. Omli, and F. MacIntyre 1989. Linneaus: Interactive taxonomy using the Macintosh computer and Hypercard. BioScience 39:635-638.

Estep, K.W., R. Sleys, and E.E. Syvertsen 1993. "Linneaus" and beyond: Workshop report on multimedia tools for the identification and database storage of biodiversity. Hydrobiologia 269/270:519-525.

Estes, J.A. and J.F. Palmisano 1974. Sea otters: Their role in structuring nearshore communities. Science 185:1058-1060.

Etter, R.J. and J.F. Grassle 1992. Patterns of species diversity in the deep sea as a function of particle size diversity. Nature 360:576-578.

Evans, P.G.H. 1987. The Natural History of Whales and Dolphins, Facts on File Publ., New York.

Everett, R.A., G.M. Ruiz, and J.T. Carlton. The effect of oyster mariculture on submerged aquatic vegetation: An experimental test in a Pacific Northwest estuary. In press, Mar. Ecol. Prog. Ser.

Falkowski, P.G., Z. Dunbinsky, L. Muscatine, and L. McCloskey 1993. Population control in symbiotic corals. BioScience 43:606-611.

Farmer, M.W., J.A. Ward, and B.E. Luckhurst 1989. Development of spiny lobster (*Panulirus argus*) phyllosoma larvae in the plankton near Bermuda. Pp. 289-301 in Proceedings of 39th Annual Gulf and Caribbean Fisheries Institute, G.T. Waugh and M.H. Goodwin, eds., Charleston, South Carolina.

Fasham, M.J.R. 1978. The statistical and mathematical analysis of plankton patchiness. Oceanogr. Mar. Biol. Ann. Rev. 16:43-80.

Fasham, M.J.R., J.L. Sarmiento, R.D. Slater, H.W. Ducklow, and R. Williams 1993. Ecosystem behavior at Bermuda Station "S" and Ocean Weather Station "India": A general circulation model and observational analysis. Global Biogeochem. Cycles 7:379-415.

Fautin, D.G., ed. 1988. Biomedical Importance of Marine Organisms. Mem. Calif. Acad. Sci. No. 13, San Francisco.

Fenchel, T.C. 1977. Aspects of the decompostion of seagrasses. Pp. 123-145 in Seagrass Ecosystems: A Scientific Perspective, C.P. McRoy and C. Helfferich, eds. Dekker, New York.

Finnerty, J.R. and B.A. Block 1992. Direct sequencing of mitochondrial DNA detects highly divergent haplotypes in blue marlin. Molec. Mar. Biol. Biotech. 1:206-214.

Fleishmann, E.M. 1989. The measurement and penetration of ultraviolet radiation into tropical marine water. Limnol. Oceanogr. 34:1623-1629.

Franks, J.S. 1992. Sink or swim: Accumulation of biomass at fronts. Mar. Ecol. Prog. Ser. 82:1-12.

Fua, L.L., W.T. Liu, and M.R. Abbott 1990. Satellite remote sensing of the ocean. Pp. 1193-1236 in The Sea, Volume 8, B. Le Mehaute and D.M. Hanes, eds. Wiley-Interscience, New York.

Fuhrman, J.A., K. McCallum, and A.A. Davis 1992. Novel major archaebacterial group from marine plankton. Nature 356:148-149.

Fuhrman, J.K., K. McCallum, and A.A. Davis 1993. Phylogenetic diversity of subsurface marine microbial communities from the Atlantic and Pacific Oceans. Appl. Environ. Microbiol. 59:1294-1302.

GESAMP 1991. The State of the Marine Environment. Blackwell Scientific Publications, Oxford Press, New York.

Gharrett, A.J. and W.W. Smoker 1993. A perspective on the adaptive importance of genetic infrastructure in salmon populations to ocean ranching in Alaska. Fish. Res. 18:45-58.

Gilpin, M. and I. Hanski, eds. 1991. Metapopulation Dynamics: Empirical and Theoretical Investigations. Academic Press, New York.

Giovannoni, S.J., T.B. Britschgi, C.L. Moyer, and K.G. Field 1990. Genetic diversity in Sargasso Sea bacterioplankton. Nature 345:60-63.

Giovannoni, S.J. and S.C. Cary 1993. Probing marine systems with ribosomal RNAs. Oceanography 6:95-104.

Givnish, T.J. 1994. Does diversity beget stability? Nature 371:113-114.

Gleason, D.F. and G.M. Wellington 1993. Ultraviolet radiation and coral bleaching. Nature 365:836-837.

Gleason, J.F., P.K. Bhartia, and J.R. Herman 1993. Record low global ozone in 1992. Science 260:253-256.

GLOBEC 1991. GLOBEC Workshop on Acoustical Technology and the Integration of Acoustical and Optical Sampling Methods. Global Ocean Ecosystems Dynamics Report No. 4.

GLOBEC 1992. Optics Technology Workshop Report. Global Ocean Ecosystems Dynamics Report No. 8.

Glynn, P.W. 1988. El Niño-Southern Oscillation 1982-83: Nearshore population, community, and ecosystem responses. Ann. Rev. Ecol. Syst. 19:309-345.

Glynn, P.W. and M.W. Colgan 1992. Sporadic disturbances in fluctuating coral reef environments: El Niño and coral reef development in the Eastern Pacific. Amer. Zool. 32:707-718.

Gobert, B. 1994. Size structures of demersal catches in a multispecies multigear tropical fishery. Fish. Res. 19:87-104.

Gooday, A.J. and C.M. Turley 1990. Responses by benthic organisms to inputs of organic material to the ocean floor: A review. Phil. Trans. Roy. Soc. Lond. A 331:119-138.

Graham, M. 1955. Effect of trawling on animals of the sea bed. Deep Sea Res. 3 (supplement):1-6.

Grassle, J.F. 1986. The ecology of deep-sea hydrothermal vent communities. Adv. Mar. Biol. 23:301-362.

Grassle, J.F. 1989. Species diversity in deep-sea communities. Trends Ecol. Evol. 4:12-15.

Grassle, J.F. 1991. Deep-sea benthic biodiversity. BioScience 41:464-469.

Grassle, J.F. and J.P. Grassle 1978. Life histories and genetic variation in marine invertebrates. Pp. 347-364 in Marine Organisms, B. Battaglia and J. Beardmore, eds., Plenum, New York.

Grassle, J.F., P. Lasserre, A.D. McIntyre, and G.C. Ray 1991. Marine biodiversity and ecosystem function. Biol. Intern. (Special Issue 23) IUBS, Paris.

Grassle, J.F. and N.J. Maciolek 1992. Deep-sea species richness: Regional and local diversity estimates from quantitative bottom samples. Amer. Nat. 139:313-341.

Grassle, J.F., N.J. Maciolek, and J.A. Blake 1990. Are deep-sea communities resilient? Pp. 386-394 in The Earth in Transition, Patterns and Processes of Biotic Impoverishment, G.M. Woodwell, ed. Cambridge University Press, New York.

Grassle, J.P. 1980. Polychaete sibling species. Pp. 25-32 in Aquatic Oligochaete Biology, R.O. Brinkhurst and D.G. Cook, eds. Plenum, New York.

Grassle, J.P., 1983. Personal communication to C.A. Butman.

Grassle, J.P. and J.F. Grassle 1976. Sibling species in the marine pollution indicator *Capitella* (Polychaeta). Science 192:567-569.

Greene, C.H. and P.H. Wiebe 1990. Bioacoustical oceanography: New tools for zooplankton and micronekton research in the 1990s. Oceanography 3:12-17.

Greene, C.H., P.H. Wiebe, and J. Burczynski 1989. Analyzing zooplankton size distributions using high-frequency sound. Limnol. Oceanogr. 34:129-139.

Greene, C.H., P.H. Wiebe, J. Burczynski, and M.J. Youngbluth 1988. Acoustical detection of high-density demersal krill layers in the submarine canyons off Georges Bank. Science 241:359-361.

Grigg, R.W. 1984. Resource management of precious corals: a review and application to shallow reef building corals. Mar. Ecol. 5:57-74.

Grigg, R.W. 1994. Science management of the world's fragile coral reefs. Coral Reefs 13:1.

Grunbaum, D. 1992. Aggregation models of individuals seeking a target density. Ph.D. thesis, Cornell University, Ithaca, New York.

Grunbaum, D. and A. Okubo 1994. Modeling animal aggregations. In Frontiers of Theoretical Biology, S.A. Levin, ed. Springer-Verlag, in press.

Hallegraeff, G.M. 1993. A review of harmful algal blooms and their apparent global increase. Phycologia 32:79-99.

Hallock, P., F.E. Muller-Karger, and J.C. Halas 1993. Coral reef decline. Res. Explor. 9: 358-378.

Hamner, W.M., P.P. Hamner, S.W. Strand, and R.W. Gilmer 1983. Behavior of Antarctic krill, *Euphausia superba*: Chemoreception, feeding, schooling, and molting. Science 220:433-435.

Hardy, J.T. and H. Gucinski 1989. Stratospheric ozone depletion: Implications for marine ecosystems. Oceanography 2:18-21.

Hargis, W.J. and D.S. Haven 1988. Rehabilitation of the troubled oyster industry of the lower Chesapeake Bay. J. Shellfish Res. 7:271-279.

Haury, L.R., J.A. McGowan, and P.H. Wiebe 1978. Patterns and processes in the time-space scales of plankton distribution. Pp. 277-327 in Spatial Pattern in Plankton Communities, J.H. Steele, ed. Plenum, New York.

Hay, M.E. 1984. Patterns of fish and urchin grazing on Caribbean coral reefs: Are previous results typical? Ecology 65:446-454.

Hodgson, G. 1989. Effects of sedimentation on Indo-Pacific reef corals. Ph.D. dissertation, Department of Zoology, University of Hawaii.

Hodgson, G. and J.A. Dixon 1988. Logging versus fisheries and tourism in Palawan. Occ. Paper 7, East-West Center, Honolulu, Hawaii.

Hofmann, E.E. 1988. Plankton dynamics on the outer southeastern U.S. continental shelf. Part III: A coupled physical-biological model. J. Mar. Res. 46:919-946.

Hofmann, E.E. and J.W. Ambler 1988. Plankton dynamics on the outer southeastern U.S. continental shelf. Part II: A time-dependent biological model. J. Mar. Res. 46:883-917.

Horoshilov, V.S. 1993. Seasonal dynamics of the Black Sea population of the ctenophore *Mnemiopsis leidyi*. Oceanography 33:558-562.

Hughes, J.M.R. and B. Goodall 1992. Marine pollution. Pp. 97-114 in Environmental Issues in the 1990s, A.M. Mannion and S.R. Bowlby, eds. John Wiley and Sons, New York.

Hunter, J.R., S.E. Kaupp, and J.H. Taylor 1981. Effects of solar and artificial ultraviolet-B radiation on larval northern anchovy, *Engraulis mordax*. Photochem. Photobiol. 34:477-486.

Hutchings, P. 1990. Review of the effects of trawling on macrobenthic epifaunal communities. Aust. J. Mar. Fresh. Res. 41:111-120.

Jackson, J.B.C. 1991. Adaptation and diversity of reef corals. BioScience 41:475-482.

Jackson, J.B.C. 1992. Pleistocene perspectives on coral reef community structure. Amer. Zool. 32:719-731.

Jackson, J.B.C. 1994. Constancy and change of life in the sea. Phil. Trans. R. Soc. Lond. B, 344:55-60.

Jackson, J.B.C., P. Jung, A.G. Coates, and L.S. Collins 1993. Diversity and extinction of tropical American mollusks and emergence of the Isthmus of Panama. Science 260:1624-1626.

Jokiel, P.L. 1980. Solar ultraviolet radiation and coral reef epifauna. Science 207:1069-1071.

Jokiel, P.L. and R.H. York, Jr. 1984. Importance of ultraviolet radiation in photoinhibition of microalgal growth. Limnol. Oceanogr. 29:192-199.

Jones, J.B. 1992. Environmental impact of trawling on the seabed: A review. N. Z. J. Mar. Fresh. Res. 26:59-67.

Jones, M.L., ed. 1985. Hydrothermal vents of the Eastern Pacific: An overview. Bull. Biol. Soc. Wash. 6, 545 pp.

Kangas, P., H. Autio, G. Hallfors, H. Luther, A. Niemi, and H. Salemaa 1982. A general model of the decline of *Fucus vesiculosus* at Tvarminne, south coast of Finland in 1977-81. Acta Bot. Fenn. 118:1-27.

Karentz, D., J.E. Cleaver, and D.L. Mitchell 1991. Cell survival characteristics and molecular responses of Antarctic phytoplankton to ultraviolet-B radiation. J. Phycol. 27:326-341.

Kato, H. 1987. Density dependent changes in growth parameters of the Southern Minke whale. Sci. Rept. Whales Res. Inst. 38:47-73.

Kautsky, N., H. Kautsky, U. Kautsky, and M. Waern 1986. Decreased depth penetration of *Fucus vesiculosus* (L.) since the 1940's indicates eutrophication of the Baltic Sea. Mar. Ecol. Prog. Ser. 28:1-8.

Kennicutt, M.C., J.M. Brooks, R.R. Bidigare, J.J. McDonald, D.L. Adkison, and S.H. Macko 1989. An upper slope 'cold' seep community: northern California, 1989. Limnol. Oceanogr. 34:635-640.

Kennish, M.J. 1992. Ecology of Estuaries: Anthropogenic Effects. CRC Press, Boca Raton, Florida, 494 pp.

Kills, U. 1992. The EcoScope and dynIMAGE: microscale tools for in situ studies of predator prey interactions. Arch. Hydrobiol. Beih. Ergebn. Limnol. 36:83-96.

Kinne, O., ed. 1984. Marine Ecology V, Ocean Management. Wiley, Chichester, England.

Knowlton, N. 1992. Thresholds and multiple stable states in coral reef community dynamics. Amer. Zool. 32:674-682.

Knowlton, N. 1993. Sibling species in the sea. Ann. Rev. Ecol. Syst. 24:189-216.

Knowlton, N. 1994. Personal communication to C.A. Butman.

Knowlton, N. and J.B.C. Jackson 1994. New taxonomy and niche partitioning on coral reefs: Jack of all trades or master of some? Trends Ecol. Evol. 9:7-9.

Knowlton, N., J.C. Lang, and B.D. Keller. 1990. Case study of natural populatin collapse: Post-hurricane prediation on Jamaican staghorn corals. Smithson. Contrib. Mar. Sci. 31:1-25.

Knowlton, N., E. Weil, L.A. Weigt, and H.M. Guzmán 1992. Sibling species in *Montastraea annularis*, coral bleaching, and the coral climate record. Science 255:330-333.

Kramer, K.J.M. 1990. Effects of increased solar UV-B radiation on coastal marine ecosystems: An overview. Pp 195-210 in Expected Effects of Climatic Change on Marine Coastal Ecosystems, 1st ed, J.J. Beukema, W.J. Wolff, and J.J.M. Brouns, eds., Kluwer Academic Publishers, Netherlands.

Kuhlmann, D.H.H. 1988. The sensitivity of coral reefs to environmental pollution. Ambio 17:13-21.

La Flamme, R.E. and R.A. Hites 1978. The global distribution of polycyclic aromatic hydrocarbons in recent sediments. Geochim. Cosmo. Acta 42:289-303.

Lannan, J.E., G.A.E. Gall, J.E. Thorpe, C.E. Nash, and B.E. Ballachey 1989. Genetic resource management of fish. Genome 31:798-804.

Larsson, U., R. Elmgren, and F. Wulff 1985. Eutrophication and the Baltic Sea: causes and consequences. Ambio 14:9-14.

Lasserre, P., A.D. McIntyre, J.C. Ogden, G.C. Ray, and J.F. Grassle, eds. 1994. Marine laboratory networks for the study of the biodiversity, function, and management of marine ecosystems. Biol. Int. Spec. Issue No. 31, 33 pp.

Launiainen, J., J. Pokki, J. Vainio, J. Niemimaa, and A. Voipio 1989. Long-term changes in the Secchi depth in the northern Baltic Sea. XIV Geofysikan Päivät Helsinki, pp. 117-121.

Laws, R.M. 1977. Seals and whales of the Southern Ocean. Phil. Trans. R. Soc. Lond. B, 279:81-96.

Laws, R.M. 1985. The ecology of the Southern Ocean. Amer. Sci. 73:26-38.

Leaman, B.M. 1991. Reproductive styles and life history variables relative to exploitation and management of *Sebastes* stocks. Envir. Biol. Fishes 30:253-271.

Lee, W.L., D.M. Devaney, W.K. Emerson, V.R. Ferris, C.W. Hart, Jr., E.N. Kozloff, F.H. Nichols, D.L. Pawson, D.F. Soule, and R.M. Woollacott 1978. Resources in invertebrate systematics. Amer. Zool. 18:167-185.

Lessios, H.A. 1988. Mass mortality of *Diadema antillarum* in the Caribbean: What have we learned? Ann. Rev. Ecol. Syst. 19:371-393.

Letelier, R.M., R.R. Bidigare, D.V. Hebel, M. Ondrusek, C.D. Winn, and D.M. Karl 1993. Temporal variability of phytoplankton community structure based on pigment analysis. Limnol. Oceanogr. 38:1420-1437.

Levin, L.A. 1990. A review of methods for labeling and tracking marine invertebrate larvae. Ophelia 32:115-144.

Levin, L.A., D. Huggett, P. Myers, T. Bridges, and J. Weaver 1993. Rare-earth tagging methods for the study of larval dispersal by marine invertebrates. Limnol. Oceanogr. 38:346-360.

Levin, S.A. 1992. The problem of pattern and scale in ecology. Ecology 73:1943-1967.

Levin, S.A., A. Morin, and T.H. Powell 1989. Patterns and processes in the distribution and dynamics of Antarctic krill. Pp. 281-299 in Scientific Committee for the Conservation of Antarctic Marine Living Resources Selected Scientific Papers, Part I, SC-CAMLR-SSP/5, CCAMLR, Hobart, Tasmania, Australia.

Lobel, P.B., S.P. Belkhode, S.E. Jackson, and H.P. Longerich 1990. Recent taxonomic discoveries concerning the mussel *Mytilus*: Implications for biomonitoring. Arch. Environ. Contam. Toxicol. 19:508-512.

Longhurst, A.R., ed. 1981. Analysis of Marine Ecosystems. Academic Press, London.

Lovejoy, T.E. 1980. A projection of species extinctions. Pp. 328-331 in Vol. 2, The Global 2000 Report to the President. Entering the 21st Century, G.O. Barney (study director), Council on Environ. Quality, U.S. Government Printing Office, Washington, D.C.

Lubchenco, J., A.M. Olson, L.B. Brubaker, S.R. Carpenter, M.M. Holland, S.P. Hubbell, S.A. Levin, J.A. MacMahon, P.A. Matson, J.M. Melillo, H.A. Mooney, C.H. Peterson, H.R. Pulliam, L.A. Real, P.J. Regal, and P.G. Risser 1991. The sustainable biosphere initiative: An ecological research agenda. Ecology 72:371-412.

MacDonald, I.R., J.F. Reilly II, N.L. Guinasso, Jr., J.M. Brooks, R.S. Carney, W.A. Bryant, and T.J. Bright 1990. Chemosynthetic mussels at a brine-filled pockmark in the northern Gulf of Mexico. Science 248:1096-1099.

Mangel, M. 1993. Models, physics and predictive biological oceanography: KNOW your organism. U.S. GLOBEC News 4:1-2.

Mannion, A.M. 1992. Acidification and eutrophication. Pp. 177-195 in Environmental Issues in the 1990s, A.M. Mannion and S.R. Bowlby, eds. John Wiley & Sons, New York.

Marine Life Resources Workshop 1989. Marine models in biomedical research. Biol. Bull. 176:337-348.

Martin, A.P., R. Humphreys, and S.R. Palumbi 1992. Population genetic structure of the armorhead, *Pseudopentaceros wheeleri*, in the North Pacific Ocean: Application of the polymerase chain reaction of fisheries problems. Can. J. Fish. Aquat. Sci. 49:2385-2391.

Martin, J.H. and 43 additional authors 1994. Testing the iron hypothesis in the ecosystems of the equatorial Pacific Ocean. Nature 371:123-129.

Matishov, G.G. and L.G. Pavlova 1994. Degradation of ecosystems of the north European seas under the effect of fishery and pathways of their recovery. Izvestiya Akademii Nauk Seriya Biologicheskaya 1:119-126.

Maurer, D., L. Watling, and R. Keck 1971. The Delaware oyster industry: A reality? Trans. Amer. Fish. Soc. 100:100-111.

May, R.M. 1988. How many species are there on earth? Science 241:1441-1449.

May, R.M. 1992. Bottoms up for the oceans. Nature 357:278-279.

May, R.M., J.R. Beddington, C.W. Clark, S.J. Holt, and R.M. Laws 1979. Management of multispecies fisheries. Science 205:267-277.

McCain, B.B., S.L. Chan, M.M. Krahn, D.W. Brown, M.S. Myers, J.T.Landahl, S. Pierce, R.C. Clark, Jr., and U. Varanasi 1992. Chemical contamination and associated fish diseases in San Diego Bay. Environ. Sci. Tech. 26:725-733.

McCleave, J.D. 1993. Physical and behavioral controls on the oceanic distribution and migration of leptocephali. J. Fish. Biol. 43 (Suppl. A):243-273.

McDonald, J.H., R. Seed, and R.K. Koehn 1992. Allozymes and morphometric characters of three species of *Mytilus* in the Northern and Southern Hemispheres. Mar. Biol. 111:323-333.

McGowan, J.A. and P.W. Walker 1993. Pelagic diversity patterns. Pp. 203-214 in Species Diversity in Ecological Communities, R.E. Ricklefs and D. Schluter, eds. Univ. of Chicago Press, Chicago.

Mileikovsky, S.A. 1971. Types of larval development in marine bottom invertebrates, their distribution and ecological significance: A re-evaluation. Mar. Biol. 10:193-213.

Miller, K.M., P. Jones, and J. Roughgarden 1991. Monoclonal antibodies as species-specific probes in oceanographic research: Examples with intertidal barnacle larvae. Mol. Mar. Biol. Biotech. 1:35-47.

Mork, J. 1991. One-generation effects of farmed fish immigration on the genetic differentiation of wild Atlantic salmon in Norway. Aquaculture 98:267-276.

Murphy, G.I. 1967. Vital statistics of the Pacific sardine (*Sardinops caerulea*) and the population consequences. Ecology 48:731-736.

Myers, M.S., J.T. Landahl, M.M. Krahn, and B.B. McCain 1991. Relationships between hepatic neoplasms and related lesions and exposure to toxic chemicals in marine fish from the U.S. west coast. Environ. Health Perspec. 90:7-15.

Nalepa, T.F. and D.W. Schloesser, eds. 1993. Zebra Mussels: Biology, Impacts and Control, Lewis Publishers, Ann Arbor, Michigan.

National Environmental Research Council 1992. Evolution and Biodiversity: The New Taxonomy, Report of the National Environmental Research Council, Great Britain.

National Oceanic and Atmospheric Administration 1992. Our Living Oceans: Report on the Status of U.S. Living Marine Resources, 1992. NOAA Tech. Memo, NMFS-F/SPO-2, Washington, D.C., 148 pp.

National Research Council 1980. International Mussel Watch. Report of a Workshop. National Academy of Sciences, Washington, D.C.

National Research Council 1985. Oil in the Sea: Impacts, Fates, and Effects. National Academy Press, Washington, D.C., 601 pp.

National Research Council 1987. Recruitment Processes and Ecosystem Structure of the Sea. A Report of a Workshop. National Academy Press, Washington, D.C.

National Research Council 1992. Oceanography in the Next Decade: Building New Partnerships. National Academy Press, Washington, D.C., 202 pp.

National Research Council 1993. A Biological Survey for the Nation. National Academy Press, Washington, D.C., 205 pp.

National Research Council 1994a. Environmental Science in the Coastal Zone: Issues for Further Research. National Academy Press, Washington, D.C., 172 pp.

National Research Council 1994b. The Ocean's Role in Global Change: Progress of Major Research Programs. National Academy Press, Washington, D.C., 85 pp.

Nee, S. and R.M. May 1992. Dynamics of metapopulations: Habitat destruction and competitive coexistence. J. Anim. Ecol. 61:37-40.

Newell, R. 1988. Ecological changes in Chesapeake Bay; are they the result of over-harvesting the American oyster *Crassostrea virginica*? Understanding the Estuary; Advances in Chesapeake Bay Research. Chesapeake Research Consortium Publication #29, Baltimore.

Newman, L.J. and L.R.G. Cannon 1994. Biodiversity of tropical polyclad flatworms from the Great Barrier Reef, Australia. Mem. Queensl. Mus. 36:159-163.

Nichols, F.H., J.K. Thompson, and L.E. Schemel 1990. Remarkable invasion of San Francisco Bay (California, USA) by the Asian clam *Potamocorbula amurensis*. II. Displacement of a former community. Mar. Ecol. Prog. Ser. 66:95-101.

Nixon, S.W., C.A. Oviatt, J. Frithsen, and B. Sullivan 1986. Nutrients and the productivity of estuarine and coastal marine ecosystems. J. Limnol. Soc. So. Afr. 12: 43-71.

Norse, E.A., ed. 1993. Global Marine Biological Diversity Strategy: Building Conservation into Decision Making. Center for Marine Conservation, Washington, D.C.

Nowell, A.R.M. and P.A. Jumars 1984. Flow environments of aquatic benthos. Ann. Rev. Ecol. Syst. 15:303-328.

Office of Technology Assessment, U.S. Congress 1993. Harmful Non-Indigenous Species in the United States. 391 pp.

Ogden, J.C. 1987. Cooperative coastal ecology at Caribbean marine laboratories. Oceanus 30:9-15.

Ogden, J.C. 1988. The influence of adjacent systems on the structure and function of coral reefs. Proc. 6th Inter. Coral Reef Symp., Australia, 1:123-129.

Ogden, J.C., J.W. Porter, N.P. Smith, A.M. Szmant, W.C. Japp, and D. Forcucci 1994. A long-term interdisciplinary study of the Florida Keys seascape. Bull. Mar. Sci. 54:1059-1071.

Olsen, G.J., C.R. Woese, and R. Overbeek 1994. The winds of evolutionary change: breathing new life into microbiology. J. Bacteriol. 176:1-6.

Olson, R.J., S.W. Chisholm, E.R. Zettler, M. Altabet, and J. Dusenberry 1990. Spatial and temporal distributions of prochlorophyte picoplankton in the North Atlantic Ocean. Deep-Sea Res. 37:1033-1051.

Olson, R.J., E.R. Zettler, S.W. Chisholm, and J.A. Dusenberry 1991. Advances in oceanography through flow cytometry. Pp. 351-399 in Particle Analysis in Oceanography, S. Demers, ed. Springer-Verlag, New York.

Olson, R.R., J.A. Runstadler, and T.D. Kocher 1991. Whose larvae? Nature 351:357-358.

Owen, R. W. 1989. Microscale and finescale variations of small plankton in coastal and pelagic environments. J. Mar. Res. 47:197-240.

Padhi, B.K. and R.K. Mandal 1994. Improper fish breeding practices and their impact on aquaculture and fish biodiversity. Current Science 66:624-626.

Paine, R.T. 1980. Food webs: Linkage, interaction strength and community infrastructure. J. Anim. Ecol. 49:667-685.

Paine, R.T. 1992. Food-web analysis through field measurement of per capita interaction strength. Nature 355:73-75.

Paine, R.T. and S.A. Levin 1981. Intertidal landscapes: Disturbance and the dynamics of pattern. Ecol. Monogr. 51:145-178.

Palumbi, S.R. 1992. Marine speciation on a small planet. Trends Ecol. Evol. 7:114-118.

Pankhurst, R.J. 1991. Practical Taxonomic Computing, Cambridge University Press.

Parrilla, G., A. Lavin, H. Bryden, M. Garcia, and R. Millard 1994. Rising temperatures in the subtropical North Atlantic Ocean over the past 35 years. Nature 369:48-51.

Pascual, M.A. and M.D. Adkison 1994. The decline of the Steller sea lion in the northeast Pacific: Demography, harvest or environment? Ecol. Appl. 4:393-403.

Perry, J.S. 1993. Understanding Our Own Planet: An Overview of Major International Scientific Activities, International Council of Scientific Unions, Paris, 36 pp.

Peterson, C.H. 1993. Implications of environmental impact analysis by application of principles derived from manipulative ecology: lessons from coastal marine case histories. Austral. J. Ecol. 18:21-52.

Peterson, M.N.A., ed. 1992. Diversity of Oceanic Life: An Evaluative Review. Center for Strategic and International Studies, Washington, D.C.

Pieper, R.R. and D. Van Holliday 1984. Acoustic measurements of zooplankton distributions in the sea. J. Cons. Int. Explor. Mer. 41:226-238.

Platt, J.R. 1964. Strong inference. Science 145:347-353.

Polovina, J.J. and S. Ralston 1986. An approach to yield assessment for unexploited resources with application to the deep slope fishes of the Marianas. Fish. Bull. 84:759-770.

Poore, G.C.B. and G.D.F. Wilson 1993. Marine species richness. Nature 361:597-599.

Porter, J.W. and O.W. Meier 1992. Quantification of loss and change in Floridian reef coral populations. Amer. Zool. 32:625-640.

Powell, T.M. and A. Okubo 1994. Turbulence, diffusion and patchiness in the sea. Phil. Trans. R. Soc. Lond. B 343:11-18.

Powers, D.A., F.W. Allendorf, and T.T. Chen 1990. Application of molecular techniques to the study of marine recruitment problems. Pp. 104-121 in Large Marine Ecosystems. Patterns,

Processes and Yields, K. Sherman, L.M. Alexander, and B.D. Gold, eds. American Association for the Advancement of Science, Washington, D.C.

Proctor, L.M. and J.A. Fuhrman 1990. Viral mortality of marine bacteria and cyanobacteria. Nature 343:60-62.

Ray, G.C. and J.F. Grassle 1991. Marine biological diversity. BioScience 41:453-457.

Ray, G.C., B.P. Hayden, A.J. Bulger, Jr., and M.G. McCormick-Ray 1992. Effects of global warming on the biodiversity of coastal-marine zones. Pp. 91-104 in Global Warming and Biological Diversity, R.L. Peters and T.E. Lovejoy, eds. Yale University Press, New Haven, Connecticut.

Rex, M.A., C.T. Stuart, R.R. Hessler, J.A. Allen, H.L. Sanders, and G.D.F. Wilson 1993. Global-scale latitudinal patterns of species diversity in the deep-sea benthos. Nature 365:636-639.

Richmond, R.H. 1993. Coral reefs: Present problems and future concerns resulting from anthropogenic disturbance. Amer. Zool. 33:524-536.

Ricklefs, R.E. 1987. Community diversity: relative roles of local and regional processes. Science 235:167-171.

Ricklefs, R.E. 1990. Scaling pattern and process in marine ecosystems. Pp. 169-178 in Large Marine Ecosystems: Patterns, Processes, and Yields, K. Sherman, L.M. Alexander, and B.D. Gold, eds. American Association for the Advancement of Science, Washington, D.C.

Ricklefs, R.E. and D. Schluter, eds. 1993. Species Diversity in Ecological Communities. Historical and Geographical Perspectives. University of Chicago Press, Chicago, 414 pp.

Ritz, D.A., M.E. Lewis, and M. Shen 1989. Response to organic enrichment of infaunal macrobenthic communities under salmonid sea cages. Mar. Biol. 103:211-214.

Robblee, M.B., T.R. Barber, P.R. Carlson, Jr., M.J. Durako, J.W. Fourqurean, L.K. Muehlstein, D. Porter, L.A. Yarbro, R.T. Zieman, and J.C. Zieman 1991. Mass mortality of the tropical seagrass *Thalassia testudinum* in Florida Bay. Mar. Ecol. Prog. Ser. 71:297-299.

Robertson, A.I. and D.M. Alongi, eds. 1992. Tropical Mangrove Ecosystems. Coastal and Estuarine Studies vol. 41. American Geophysical Union, Washington, D.C., 329 pp.

Robertson, D.A. and P.J. Grimes 1983. The New Zealand orange roughy fishery. Pp. 15-20 in New Zealand Finfish Fisheries: The Resources and Their Management, J.L. Taylor, and G.G. Baird, eds. Trade Publs. Ltd., Auckland, New Zealand.

Rogers, C.S. 1985. Degradation of Caribbean and western Atlantic coral reefs and decline of associated reef fisheries. Proc. 5th Intl. Coral Reef Symp., Tahiti, Vol. 6:491-496.

Rönnberg, O., K. Ådjers, C. Ruokolahti, and M. Bondestam 1992. Effects of fish farming on growth, epiphytes and nutrient content of *Fucus vesiculosus* L. in Åland Archipelago, northern Baltic Sea. Aquat. Bot. 42:109-120.

Rosel, P.E., A.E. Dizon, and J.E. Heyning 1994. Genetic analysis of sympatric morphotypes of common dolphins (genus *Delphinus*). Mar. Biol. 119:159-167.

Rothlisberg, P.C., J.A. Church, and A.M.G. Forbes 1983. Modelling the advection of vertically migrating shrimp larvae. J. Mar. Res. 41:511-538.

Roughgarden, J., S. Gaines, and H. Possingham 1988. Recruitment dynamics in complex life cycles. Science 241:1460-1466.

Rowan, R. and N. Knowlton. Intraspecific diversity and ecological zonation in coral-algal symbiosis. In press, Proc. Natl. Acad. Sci. USA.

Rowan, R. and D.A. Powers 1991. A molecular genetic classification of zooxanthellae and the evolution of animal-algal symbioses. Science USA 251:1348-1351.

Rowan, R. and D.A. Powers 1992. Ribosomal RNA sequences and the diversity of symbiotic dinoflagellates (zooxanthellae). Proc. Natl. Acad. Sci. 89:3639-3643.

Ruppert, E.E. and R.S. Fox 1988. Seashore Animals of the Southeast. A Guide to Common Shallow-Water Invertebrates of the Southeastern Atlantic Coast. University of South Carolina Press, 429 pp.

Russ, G.R. 1991. Coral reef fisheries: Effects and yields. Pp. 601-636 in The Ecology of Fishes on Coral Reefs, P.F. Sale, ed. Academic Press, New York.

Ryman, N. 1991. Conservation genetics considerations in fishery management. J. Fish. Biol. 39: 211-234.

SA2000 1994. Systematics Agenda 2000: Charting the Biosphere. Technical Report. Produced by Systematics Agenda 2000. Dept. Ornithology, American Museum of Natural History, New York, New York and Herbarium, New York Botanical Gardens, Bronx, New York.

Safina, C. 1993. Bluefin tuna in the West Atlantic: Negligent management and the making of an endangered species. Conser. Biol. 7:229-234.

Salvat, B., ed. 1987. Human Impacts on Coral Reefs: Facts and Recommendations. Antenne Museum Ecole Practique des Hautes Etudes, French Polynesia.

Schaff, T., L. Levin, N. Blair, D. DeMaster, R. Pope, and S. Boehme 1992. Spatial heterogeneity of benthos on the Carolina continental slope: Large (100 km)-scale variation. Mar. Ecol. Prog. Ser. 88:143-160.

Schalk, P.H. 1994. Information and identification systems for marine organisms on CD-ROM. The Plankton Newsletter, Jan. 1994, pp. 6-11.

Scheltema, R. S. 1986. On dispersal and planktonic larvae of benthic invertebrates: An eclectic overview and summary of problems. Bull. Mar. Sci. 39:290-322.

Schmidt, T.M., E.F. DeLong, and N.R. Pace 1991. Analysis of a marine picoplankton community by 16S rRNA gene cloning and sequencing. J. Bacteriol. 173:4371-4378.

Schneider, C.W. and R.B. Searles 1991. Seaweeds of the Southeastern United States: Cape Hatteras to Cape Canaveral. Duke University Press, Durham, North Carolina, 553 pp.

Schubel, J.R. 1994. Coastal pollution and waste management. Pp. 124-148 in Environmental Science in the Coastal Zone: Issues for Further Research. Proceedings of a retreat held by the National Research Council, National Academy Press, Washington, D.C.

Sebens, K.P. 1994. Biodiversity of coral reefs: What are we losing and why? Amer. Zool. 34:115-133.

Shapiro, D.Y. 1987. Reproduction in groupers. Pp. 295-327 in Tropical Snappers and Groupers: Biology and Fisheries Management, J.J. Polovina and S. Ralston, eds. Westview Press, Boulder, Colorado.

Sherman, K., L.M. Alexander, and B.D. Gold, eds. 1990. Large Marine Ecosystems. Patterns, Processes, and Yields. American Association for the Advancement of Science, Washington, D.C., 242 pp.

Shushkina, E.A., G.G. Nikolaeva, T.A. Lukasheva 1990. Changes in the structure of the Black Sea planktonic community at mass reproduction of sea gooseberries Mnemiopsis leidyi (Agassiz). Oceanology 51:54-60.

Sieburth, J. McN., P W. Johnson, and P.E. Hargraves 1988. Ultrastructure and ecology of Aureococcus anophagefferens, gen. et sp. nov. (Chrysophyceae): The dominant picoplankter during a bloom in Narragansett Bay. J. Phycol. 24:416-425.

Siegfried, W.R., ed. 1994. Rocky Shores: Exploitation in Chile and South Africa. Springer-Verlag, Berlin, 190 pp.

Sieracki, M.E. and C.L. Viles 1990. Color image-analyzed fluorescence microscopy: A new tool for marine microbial ecology. Oceanography 3:30-36.

Skreslet, S., ed. 1986. The Role of Freshwater Outflow in Coastal Marine Ecosystems. NATO ASI Series. Springer-Verlag, New York.

Smayda, T.J. and Y. Shimizu, eds. 1993. Toxic Phytoplankton Blooms in the Sea. Elsevier, London, 952 pp.

Smith, C.R. 1985. Food for the deep sea: Utilization, dispersal, and flux of nekton falls at the Santa Catalina Basin floor. Deep-Sea Res. 32:417-442.

Smith, C.R. 1992. Whale falls: chemosynthesis on the deep seafloor. Oceanus 36:74-78.

Smith, C.R. and R.R. Hessler 1987. Colonization and succession in deep-sea ecosystems. Trends Ecol. Evol. 2:359-363.

Smith, C.R., P.A. Jumars, and D.D. DeMaster 1986. In situ studies of megafaunal mounds indicate rapid sediment turnover and community response at the deep-sea floor. Nature 323:251-253.

Smith, C.R., H. Kukert, R.A. Wheatcroft, P.A. Jumars, and J.W. Deming 1989. Vent fauna on whale remains. Nature 341:27-28.

Smith, P.E., M.D. Ohman, and L.E. Eber 1989. Analysis of the patterns of distribution of zooplankton aggregations from an acoustic Doppler current profiler. CalCOFI Rept. 30:88-103.

Smith, P.J., R.I.C.C. Francis, and M. McVeagh 1991. Loss of genetic diversity due to fishing pressure. Fish. Res. 10: 309-316.

Smith, R.C., B.B. Prézlin, K.S. Baker, R.R. Bidigare, N.P. Boucher, T. Coley, D. Karentz, S. MacIntyre, H.A. Matlick, D. Menzies, M. Ondrusek, Z. Wan, and K.J. Walters 1992. Ozone depletion: Ultraviolet radiation and phytoplankton biology in Antarctic waters. Science 255: 952-959.

Smith, S.V. and R.W. Buddemeier 1992. Global change and coral reef ecosystems. Ann. Rev. Ecol. Syst. 23:89-118.

Smith, S.V., W.J. Kimmerer, E.A. Laws, R.E. Brock, and T.W. Walsh 1981. Kaneohe Bay sewage diversion experiment: Perspectives on ecosystem responses to nutritional perturbation. Pac. Sci. 35:279-396.

Smith, T.D. 1983. Changes in size of three dolphin (*Stenella* spp.) populations in the eastern tropical Pacific. Fish. Bull. 81:1-13.

Snelgrove, P.V.R., J.F. Grassle, and R.F. Petrecca 1994. Macrofaunal response to artificial enrichments and depressions in a deep-sea habitat. J. Mar. Res. 52:345-369.

Solbrig, O.T., ed. 1991. From Genes to Ecosystems: A Research Agenda for Biodiversity. International Union of Biological Sciences, Paris.

Sousa, W.P. 1985. Disturbance and patch dynamics on rocky intertidal shores. Pp. 101-124 in The Ecology of Natural Disturbance and Patch Dynamics, S.T.A. Picket and P.S. White, eds. Academic Press, Orlando.

Southward, A.J. 1989. Animal communities fuelled by chemosynthesis: life at hydrothermal vents, cold seeps and in reducing sediments. J. Zool. 217:705-709.

Steele, J.H. 1985. A comparison of terrestrial and marine ecological systems. Nature 313:355-358.

Steele, J.H., S. Carpenter, J. Cohen, P. Dayton, and R. Ricklefs 1989. Comparison of Terrestrial and Marine Ecological Systems. Report of a Workshop held in Santa Fe, New Mexico, 14 pp.

Steinbeck, J. 1954. Sweet Thursday. Viking Press, Inc., New York, 273 pp.

Sterrer, W., ed. 1986. Marine Fauna and Flora of Bermuda. A Systematic Guide to the Identification of Marine Organisms. John Wiley & Sons, New York, 742 pp.

Stuessey, T.F. and K.S. Thomson, eds. 1981. Trends, Priorities and Needs in Systematic Biology. A report to the Systematic Biology Program of the National Science Foundation, Association of Systematics Collections, Lawrence, Kansas, 51 pp.

Stockton, W.L. and T.E. DeLaca 1982. Food falls in the deep sea: Occurrence, quality and significance. Deep-Sea Res. 29:157-169.

Strathmann, R.R. 1985. Feeding and nonfeeding larval development and life-history evolution in marine invertebrates. Ann. Rev. Ecol. Syst. 16:339-361.

Strathmann, R.R. 1990. Why life histories evolve differently in the sea. Amer. Zool. 30:197-207.

Suchanek, T.H. 1993. Oil impacts on marine invertebrate populations and communities. Amer. Zool. 33:510-523.

Suchanek, T.H. 1994. Temperate coastal marine communities: Biodiversity and threats. Amer. Zool. 34:100-114.

Suchanek, T.H., S.L. Williams, J.C. Ogden, D.K. Hubbard, and I.P. Gill 1985. Utilization of shallow-water seagrass detritus by Caribbean deep-sea macrofauna: Delta^{13} C evidence. Deep-Sea Res. 32:201-214.

Takada, H., J.W. Farrington, M.H. Bothner, C.G. Johnson, and B.W. Tripp 1994. Transport of sludge-derived organic pollutants to deep-sea sediments at Deep Water Dump Site 106. Environ. Sci. Technol. 28:1062-1072.

Teal, J. and M. Teal 1969. Life and Death of the Salt Marsh. Audubon/Ballantine, New York.

Thayer, G.W., ed. 1992. Restoring the Nation's Marine Environment. Maryland Sea Grant, College Park, 716 pp.

Thayer, G.W., K.A. Bjorndal, J.C. Ogden, S.L. Williams, and J.C. Zieman 1984. Role of larger herbivores in seagrass communities. Estuaries 7:351-376.

Thistle, D. and J.E. Eckman 1990. The effect of a biologically produced structure on the benthic copepods of a deep-sea site. Deep-Sea Res. 37:541-554.

Thorson, G. 1950. Reproductive and larval ecology of marine bottom invertebrates. Biol. Rev. 25:1-45.

Tilman, D. and J.A. Downing 1994. Biodiversity and stability in grasslands. Nature 367:363-365.

Tilman, D., J.A. Downing, and D.A. Wedin 1994. Does diversity beget stability? Tilman et al. Reply. Nature 371:114.

Tilman, D. and S. Pacala 1993. The maintenance of species richness in plant communities. Pp. 13-25 in Species Diversity in Ecological Communities, R.E. Ricklefs and D. Schluter, eds. University of Chicago Press, Chicago.

Tremblay, M.J., J.W. Loder, F.E. Werner, C. Naimie, F.H. Page, and M.M. Sinclair 1994. Drift of scallop larvae on Georges Bank: A model study of the roles of mean advection, larval behavior and larval origin. Deep-Sea Res. 41:7-49.

Tsutsumi, H., T. Kikuchi, M. Tanaka, T. Higashi, K. Imasaka, and M. Miyazaki 1991. Benthic faunal succession in a cove organically polluted by fish farming. Mar. Poll. Bull. 23:233-238.

Tunnicliffe, V. 1991. The biology of hydrothermal vents: Ecology and evolution. Oceanogr. Mar. Biol. Ann. Rev. 29:319-407.

Turner, R.D. 1973. Wood-boring bivalves, opportunistic species in the deep sea. Science 180:1377-1379.

Turner, R.D. 1981. "Wood islands" and "thermal vents" as centers of diverse communities in the deep sea. Biologiya Morya 1:3-10.

Turner, R.E. and N.N. Rabalais 1994. Coastal eutrophication near the Mississippi River delta. Science 368:619-621.

Underwood, A.J. and P.S. Petraitis 1993. Structure of intertidal assemblages in different locations: How can local processes be compared? Pp. 39-51 in Species Diversity in Ecological Communities, R.E. Ricklefs and D. Schluter, eds. University of Chicago Press, Chicago.

USGS 1994. The National Marine and Coastal Geology Program. A plan developed by the U.S. Geological Survey at the request of the U.S. Congress, U.S. Department of the Interior, U.S. Geological Survey, June 1994, 56 pp.

Valentine, J.W. and D. Jablonski 1993. Fossil communities: Compositional variation at many time scales. Pp. 341-349 in Species Diversity in Ecological Communities, R.E. Ricklefs and D. Schluter, eds. University of Chicago Press, Chicago.

VanBlaricom, G.R. and J.A. Estes 1987. The Community Ecology of Sea Otters. Springer-Verlag, N.Y.

van der Veer, H.W. 1989. Eutrophication and mussel culture in the western Dutch Wadden Sea: Impact on the benthic ecosystem — A hypothesis. Helgol. Wiss. Meeresunters. 43:517-527.

van Oppen, M.J.H., O.E. Dickmann, C. Wiencke, W.T. Stam, and J.L. Olsen 1994. Tracking dispersal routes: phytogeography of the Arctic-Antarctic disjunct seaweed *Acrosiphonia arcta* (chlorophyta). J. Phycol. 30:67-80.

Vermeij, G.J. 1987. Evolution and Escalation. Princeton University Press, Princeton, N.J.

Vermeij, G.J. 1991a. Anatomy of an invasion: The trans-Arctic interchange. Paleobiology 17:281-307.

Vermeij, G.J. 1991b. When biotas meet: Understanding biotic interchange. Science 253:1099-1104.

Vermeij, G.J. 1993. Biogeography of recently extinct marine species: Implications for conservation. Conserv. Biol. 7:391-397.

Vethaak, A.D. and T. Rheinaldt 1992. Fish disease as a monitor for marine pollution: The case of the North Sea. Rev. Fish Biol. Fish. 2:1-32.

Vogel, S. 1981. Life in Moving Fluids. The Physical Biology of Flow. Princeton University Press, Princeton, N.J., 352 pp.

Vogt, H. and W. Schramm 1991. Conspicuous decline of *Fucus* in Kiel Bay (western Baltic): What are the causes? Mar. Ecol. Prog. Ser. 69:189-194.

Von Alt, C.J. and J.F. Grassle 1992. LEO-15, an unmanned long-term environmental observatory. Proc. Oceans '92, pp. 849-854.

Walters, C.J. and R. Hilborn 1978. Ecological optimization and adaptive management. Ann. Rev. Ecol. Syst. 9:157-188.

Ward, B.B. 1990. Immunology in biological oceanography and marine ecology. Oceanography 3:30-35.

Ward, B.B. and A.F. Carlucci 1985. Marine ammonia and nitrite-oxidizing bacteria serological diversity determined by immunofluorescence in culture and in the environment. Appl. Environ. Microbiol. 50:194-201.

Ward, D.M., R. Weller, and M.B. Bateson 1990. 16s rRNA sequences reveal numerous uncultured microorganisms in a natural community. Nature 345:63-65.

Watling, L. and R. Langton 1994. Fishing, habitat disruption, and biodiversity loss. EOS, Trans. Amer. Geophys. Union 75 (3):210.

Webb, T. III and P.J. Bartlein 1992. Global changes during the last three million years: climatic controls and biotic responses. Ann. Rev. Ecol. Syst. 23:141-173.

Weber, P. 1993. Abandoned Seas. Reversing the Decline of the Oceans. Worldwatch Paper 116, Worldwatch Institute, Washington, D.C., 66 pp.

Weber, P. 1994. Net loss: fish, jobs and the marine environment. Worldwatch Paper 120, Worldwatch Institute, Washington, D.C., 76 pp.

Weinstein, M.P. 1979. Shallow marsh habitats as primary nurseries for fishes and shellfish, Cape Fear River, North Carolina. Fish. Bull. 77:339-358.

Werner, F.E., F.H. Page, D.R. Lynch, J.W. Loder, R.G. Lough, R.I. Perry, D.A. Greenberg, and M.M. Sinclair 1993. Influence of mean 3-D advection and simple behavior on the distribution of cod and haddock early life stages on Georges Bank. Fish. Oceanogr. 2:43-64.

Werner, I. and J.T. Hollibaugh 1993. *Potamocorbula amurensis*: Comparison of clearance rates and assimilation efficiencies for phytoplankton and bacterioplankton. Limnol. Oceanogr. 38:949-964.

Williams, A.B. 1988. New marine decapod crustaceans from waters influenced by hydrothermal discharge, brine, and hydrocarbon seepage. Fish. Bull. 86:263-287.

Williams, A.B. and C.L. Van Dover 1983. A new species of *Munidopsis* from submarine thermal vents of the East Pacific Rise at 21° N (Anomura: Galatheidae). Proc. Biol. Soc. Wash. 96:481-488.

Wilson, D.S. 1992. Complex interactions in metacommunities, with implications for biodiversity and higher levels of selection. Ecology 73:1984-2000.

Wilson, E.O. 1992. The Diversity of Life. Harvard University Press, Cambridge, Massachusetts.

Witbaard, R. and R. Klein 1994. Long-term trends on the effects of the southern North Sea beamtrawl fishery on the bivalve mollusc *Arctica islandica* L. (Mollusca, Bivalvia). ICES J. Mar. Sci. 51:99-105.

Witman, J.D. and K.P. Sebens 1992. Regional variation in fish predation intensity: A historical perspective in the Gulf of Maine. Oecologia 90:305-315.

Wolff, T. 1979. Macrofaunal utilization of plant remains in the deep sea. Sarsia 64:117-136.

Wood, A.M. and T. Leathem 1992. The species concept in phytoplankton ecology. J. Phycol. 28:723-729.

World Conservation Monitoring Center 1992. Global Biodiversity: Status of the Earth's Living Resources. Chapman and Hall, London.

Yentsch, C.M., F.C. Mague, and P.K. Horan, eds. 1988. Immunochemical Approaches to Coastal, Estuarine and Oceanographic Questions. Lecture Notes on Coastal and Estuarine Studies 25, Springer-Verlag, New York.

Yoder, J.A., L.P. Atkinson, T.N. Lee, H.H. Kim, and C.R. McClain 1981. Role of Gulf Stream frontal eddies in forming phytoplankton patches on the outer southeastern shelf. Limnol. Oceanogr. 26:1103-1111.

Zedler, J.B. 1994. Coastal wetlands: Multiple management problems in southern California. Pp. 107-123 in Environmental Science in the Coastal Zone: Issues for Further Research. Proceedings of a retreat held by the National Research Council, National Academy Press, Washington, D.C.

APPENDIXES

Workshop Agenda

BIOLOGICAL DIVERSITY IN MARINE SYSTEMS

May 24–26, 1994
NAS/NAE Arnold and Mabel Beckman Center
Irvine, CA

Tuesday, May 24, 1994

Remarks from Co-Chairs and National Research Council Ocean Studies
Board and the Board on Biology

Introduction of Participants

Presentation of Workshop Structure
The Focus (S. Palumbi)
Environmental Questions/Regional-Scale Approaches (J. Jackson)
Issues in Taxonomy (L. Watling)

Open Discussion

Charge to Working Groups (C.A. Butman and J. Carlton)

*Working Groups: **(A) Identification of Critical Environmental Issues**

Reconvene: Working Group Reports (10 minutes each)

Open Discussion

Working Group A: Synthesis and Discussion (C.A. Butman and J. Carlton)

*Working Groups: **(B) Identify Representative Regional-Scale Systems**

Reconvene: Working Group Reports

Open Discussion and Synthesis

Adjourn

---

*Three working groups of 16 scientists each.

Wednesday, May 25, 1994

Working Group B: Synthesis and Discussion (C.A. Butman and J. Carlton)

*Working Groups: **(C) Specific Biodiversity Research Questions**

Reconvene: Working Group Reports

Open Discussion

Working Group C: Synthesis and Discussion (C.A. Butman and J. Carlton)

†Concurrent Working Groups: **(D) Taxonomy, (E) Methods and Techniques**

Reconvene: Working Group Reports

Open Discussion

Thursday, May 26, 1994

Working Groups D and E: Synthesis and Discussion (C.A. Butman and J. Carlton)

†Concurrent Working Groups: **(F) Logistics and Coordination, (G) Products and Information Dissemination**

Reconvene: Working Group Reports

Open Discussion

Presentation and Discussion of Research Initiative (includes synthesis of Working Groups F and G)

Federal Agency Perspectives

Closing Remarks (Co-Chairs and National Research Council Ocean Studies Board/Board on Biology)

Adjourn Workshop

---

*Three working groups of 16 scientists each.
†Four concurrent working groups of 12 scientists each.

Workshop Participants

BIOLOGICAL DIVERSITY IN MARINE SYSTEMS

May 24–26, 1994
NAS/NAE Arnold and Mabel Beckman Center
Irvine, CA

Committee Members

Cheryl Ann Butman, *Co-Chair*, Woods Hole Oceanographic Institution
James Carlton, *Co-Chair*, Williams College—Mystic Seaport
George Boehlert, NOAA/NMFS, Southwest Fisheries Science Center
Susan Brawley, University of Maine
Edward DeLong, University of California—Santa Barbara
J. Frederick Grassle, Rutgers University
Jeremy Jackson, Smithsonian Tropical Research Institute
Simon Levin,* Princeton University
Arthur Nowell,* University of Washington
Robert Paine, University of Washington
Stephen Palumbi, University of Hawaii
Geerat Vermeij,* University of California—Davis
Les Watling, University of Maine

Scientific Participants

Mark Abbott, Oregon State University
Robert Andersen, Bigelow Laboratory for Ocean Sciences
Karl Banse, University of Washington
Mark Bertness, Brown University
Martin Buzas, National Museum of Natural History, Smithsonian Institution
Robert Carney, Louisiana State University/Coastal Ecology Institute

---

*Unable to attend workshop

Sallie (Penny) Chisholm, Massachusetts Institute of Technology
Jeremy Collie, University of Rhode Island
Dan Costa, University of California—Santa Cruz
Mike Dagg, Louisiana Universities Marine Consortium
Jonathan Geller, University of North Carolina—Wilmington
W. Rockwell (Rocky) Geyer, Woods Hole Oceanographic Institution
Richard Grosberg, University of California—Davis
Loren Haury, Scripps Institution of Oceanography
Mark Hay, University of North Carolina—Chapel Hill
James Hollibaugh, Tiburon Center—San Francisco State University
Nancy Knowlton, Smithsonian Tropical Research Institute
Thomas Kocher, University of Delaware
Rikk Kvitek, Moss Landing Marine Laboratories
Jane Lubchenco, Oregon State University
Laurence Madin, Woods Hole Oceanographic Institution
Mark Ohman, Scripps Institution of Oceanography
Jeanine Olsen, University of Groningen, The Netherlands
Charles (Pete) Peterson, University of North Carolina—Chapel Hill
Thomas Powell, University of California—Davis
James Quinn, University of California—Davis
Leslie Rosenfeld, Monterey Bay Aquarium Research Institute
Carl Safina, National Audubon Society
Thomas Schmidt, Michigan State University
Theodore Smayda, University of Rhode Island
Diane Stoecker, University of Maryland—Horn Point Laboratories
David Thistle, Florida State University
Elizabeth Venrick, Scripps Institution of Oceanography
Susan Williams, San Diego State University
Anne Michelle Wood, University of Oregon

Federal Agency Representatives

Randall Alberte, Office of Naval Research
Peter Barile, National Science Foundation
Roger Griffis, NOAA, Office of the Chief Scientist
Aleta Hohn, NOAA, National Marine Fisheries Service
Steve Jameson,* NOAA, Office of Resources, Conservation, and Assessment
Michael Sissenwine, NOAA, National Marine Fisheries Service
Phillip Taylor, National Science Foundation
Donna Turgeon, NOAA, Office of Resources, Conservation, and Assessment
Donna Wieting, NOAA, Office of the Chief Scientist

---

*Also representing the National Biological Service.

Scientific Press

Marguerite Holloway, Scientific American

NRC

Eric Fischer, Board on Biology
Morgan Gopnik, Commission on Geosciences, Environment, and Resources
Mary Hope Katsouros, Ocean Studies Board
LaVoncyé Mallory, Ocean Studies Board
Mary Pechacek, Ocean Studies Board
David Wilmot, Ocean Studies Board

Acronyms

AID	Agency for International Development
AMLC	Association of Marine Laboratories of the Caribbean
ASC	Association of Systematics Collections
BB	Board on Biology
CalCOFI	California Cooperative Ocean Fisheries Investigations
CARICOMP	Caribbean Coastal Marine Productivity
CENR	Committee on Environment and Natural Resources
EPA	Environmental Protection Agency
GESAMP	Group of Experts on the Scientific Aspects of Marine Pollution
GIS	Geographical Information Systems
GLOBEC	Global Ocean Ecosystem Dynamics
ICSU	International Council of Scientific Unions
IUBS	International Union of Biological Sciences
IUMS	International Union of Microbiological Sciences
JGOFS	Joint Global Ocean Flux Study
MARS	Marine Research Stations Network
MASZP	Moored, Automated, Serial Zooplankton Pump
MGI	Microbial Genome Initiative
NAML	National Association of Marine Laboratories
NAS	National Academy of Sciences
NBS	National Biological Service
NIH	National Institutes of Health
NMFS	National Marine Fisheries Service

NOAA	National Oceanic and Atmospheric Administration
NRC	National Research Council
NSF	National Science Foundation
NSTC	National Science and Technology Council
OMP	Ocean Margins Program
OSB	Ocean Studies Board
PAHs	polycyclic aromatic hydrocarbons
PCBs	polychlorinated biphenols
PCR	polymerase chain reaction
RSVP	Rapid Sampling Vertical Profiler
SAML	Southern Association of Marine Laboratories
SCOPE	Scientific Committee on Problems of the Environment
UCAR	University Corporation for Atmospheric Research
UNESCO	United Nations Educational, Scientific, and Cultural Organization
VPR	Video Plankton Recorder
WCMC	World Conservation Monitoring Center

Glossary

algal symbiont   *see* zooxanthellae

anadromous   referring to the annual migratory behavior of adult fish (such as salmon, shad, striped bass, and lamprey) from the ocean into freshwater rivers and lakes in order to spawn

anoxia   the absence of oxygen in water and sediments

benthic   living on or in the bottom (in contrast to pelagic)

biota   all of the living organisms (plants, animals, protists, fungi, and so on) in a given region

chemoautotrophic   referring to the ability to obtain energy through chemosynthesis, i.e., the oxidation of simple compounds (oxidation is a chemical reaction in which oxygen is gained, or hydrogen or electrons are lost, from a compound)

dinoflagellate   a microscopic plant, characterized by having two lash-like structures (flagella) used for locomotion, often abundant in the open ocean; many produce light and are one of the primary contributors to bioluminescence in the ocean. Some dinoflagellates, known as zooxanthellae, are symbiotic in the tissues of corals and other tropical organisms

epipelagic   referring to the top 200 meters of the ocean

eutrophication   nutrient enrichment, typically in the form of nitrates and phosphates, often from human sources such as agriculture, sewage, and urban runoff

gyre   a circular system of water movement

hydrothermal vent   an opening in the deep-sea floor out of which rises water that has been heated by contact with molten rock; this water is often rich in dissolved compounds, such as hydrogen sulfide, which are the primary source of energy for the chemoautotrophic bacteria that form the base of vent community food webs

hypoxia low concentrations of oxygen in water and sediments

littoral the ocean shore, including the rocky intertidal, sandy beaches, and salt marshes

mariculture the growing of marine animals and plants under specialized culture conditions

mesopelagic referring to depths between 200 to 1,000 meters in the ocean

nekton swimming organisms that are able to move independently of water currents (as opposed to plankton). These include most fish, mammals, turtles, sea snakes, and aquatic birds.

oligonucleotide a few nucleotides joined together; a nucleotide is a compound formed of one molecule each of a sugar, of phosphoric acid, and of a base containing nitrogen. The nucleic acids DNA and RNA are molecules made of a large number of nucleotides.

oligotrophic low in nutrients and in primary production

pelagic living in the water column (in contrast to benthic)

photosynthesis chemical reactions in plants and plant-like organisms whereby the sun's energy is absorbed by the green pigment chlorophyll, permitting carbon dioxide and water to be synthesized into carbohydrates accompanied by the release of water and oxygen

picoplankton planktonic organisms ranging in size from 0.2 to 2.0 micrometers (there are 1,000 micrometers in one millimeter, and 25.4 millimeters in one inch)

plankton floating and drifting organisms that have limited swimming abilities and that are carried largely passively with water currents (as opposed to nekton). These include bacteria (bacterioplankton), plants and plant-like organisms (phytoplankton) and the animals (zooplankton) that eat them.

polythetic classification of organisms based upon a combination of a large number of characteristics, not all of which are possessed by every member of the group

prochlorophyte bacteria that are the smallest photosynthetic cells (less than one micrometer; see picoplankton) in the open ocean; nearly ubiquitous in the sea

prokaryote an organism whose DNA is a strand within the cell, and is not contained within a nucleus; bacteria and blue-green algae are prokaryotes

propagule a dispersal stage of a plant or animal, such as fertilized eggs, larvae, or seeds

protogynous hermaphroditism a sexual condition in which female organs or gametes develop first, followed by the development of male organs or gametes, in the same individual

stability   the ability of a given assemblage of organisms to withstand disturbance without a major change in the number of species or individuals

trophic   referring to the nutrients available to and used within a population,
community, or ecosystem

ultraviolet radiation   radiation beyond the violet (high energy) end of the vis-
ible light spectrum. UV-B is the middle range wave-length of the three UV
bands, and is largely absorbed in the Earth's atmospheric ozone layer; pro-
longed exposure to UV-B can be biologically damaging.

zooxanthellae   symbiotic dinoflagellates in corals and other organisms

Index